Praise for
Nabokov's Favorite Word Is Mauve

"A *hell* of a lot of fun. . . . There's something cheeky in the way Blatt throws genre bestsellers into his statistical blender alongside literary lions and hits puree, looking for patterns of style shared by, say, James Joyce and James Patterson."

—NPR

"Brilliant."

—*The Boston Globe*

"Nate Silver–esque number crunching meets the canon in this quirky, arresting deconstruction of literature's greatest hits."

—*O, The Oprah Magazine*

"Delivers a statistical study of literature in the vein of *Freakonomics*. . . . [Blatt] approaches the subject with the right mix of humor, hand-holding, and literary love. . . . Yield[s] insights that would be impossible to recognize on their own."

—*Paste* Magazine

"Terrific. I recommend it heartily."

—*Forbes*

"Lively . . . worthwhile. . . . Read this book thoughtfully. It's fun. And, I think, the shape of some very interesting things to come."

—*The Times* (London)

"A thoroughly entertaining romp."

—*The Millions*

"Enticingly heretical. . . . The lessons here are valuable [and] Blatt's research on diction and gender is especially revelatory. . . . His discoveries are startlingly apt."

—*The New Yorker*, Page Turner column

"Blatt takes a by-the-numbers look at literary classics and finds some fascinating patterns . . . makes a strong argument."

—Smithsonian.com

"A super fun book for lit nerds . . . [a] wonderful addition to any book-lovers' TBR pile."

—*Literary Hub* (one of the *Lit Hub*'s "Best Books About Books")

"A statistician uses curiosity and big data to uncover answers to persistent literary questions. . . . The result is a lighthearted numerical examination of words that is informative, surprising, and funny."

—*Shelf Awareness*

"Blatt doesn't just shine a light on writing, he lets in a whole new area of the electromagnetic spectrum. . . . [Blatt] has achieved something impressive with this book. I've read a lot of books about words, but none like this. . . . Anyone interested in literature or becoming a better writer will find something to like here."

—Mark Peters, *Dog Eared* blog

"Book-lovers will delight in *Nabokov's Favorite Word Is Mauve* . . . accessible, entertaining, and enlightening."

—*Bustle*

"This is really the most delicious kind of rabbit hole. . . . If you're a writer, you won't be able to resist it. If you know a writer, give this as a gift and find yourself adored. . . . It can be dipped into like a squirrel's nut hoard, enjoyed a quick nibble at a time, or dived into headfirst, one fascinating tidbit leading to the next to the next to the next."

—*Publishers Weekly*, Shelf Talker column

"Illuminating entertainment."

—*Kirkus Reviews*

"Amiable and intelligent . . . literature enthusiasts will enjoy the hypotheses [Blatt] poses and his imaginative methods."

—*Publishers Weekly*

"What fun this is! Ben Blatt's charming book applies numerical know-how to questions of literary style, teasing out insights about cliffhangers, adverbs, and whether Americans write 'more loudly' than the British. (Spoiler: WE DO!!!)"

—Jordan Ellenberg, author of *How Not to Be Wrong*

"It was statisticians, rather than historians, who cracked the centuries-old mystery of the Federalist Papers—and they did it with mere paper and pencil. Operating in the same investigative spirit—and with the benefit of vastly more powerful tools— Ben Blatt probes the literary canon for unexpected revelations and insights. The result is a literary detective story: fast-paced, thought-provoking, and intriguing."

—Brian Christian, co-author of *Algorithms to Live By*

"Ben Blatt's delightful book gives us an original big data perspective on great writers' work. Its humor, insights, and statistical displays are fascinating to behold, even as it helps us develop our own writing."

—Carl N. Morris, Professor Emeritus of Statistics, Harvard University

ALSO BY BEN BLATT

*I Don't Care If We Never Get Back: 30 Games in
30 Days on the Best Worst Baseball Road Trip Ever*
(co-written with Eric Brewster)

Nabokov's Favorite Word Is *Mauve*

What the Numbers Reveal About the Classics, Bestsellers, and Our Own Writing

Ben Blatt

Simon & Schuster Paperbacks

NEW YORK · LONDON · TORONTO · SYDNEY · NEW DELHI

Simon & Schuster Paperbacks
An Imprint of Simon & Schuster, Inc.
1230 Avenue of the Americas
New York, NY 10020

First Simon & Schuster trade paperback edition March 2018

SIMON & SCHUSTER PAPERBACKS and colophon are
registered trademarks of Simon & Schuster, Inc.

For information about special discounts for bulk purchases,
please contact Simon & Schuster Special Sales at 1-866-506-1949
or business@simonandschuster.com.

The Simon & Schuster Speakers Bureau can bring authors to your
live event. For more information or to book an event, contact the
Simon & Schuster Speakers Bureau at 1-866-248-3049
or visit our website at www.simonspeakers.com.

Interior design by Paul Dippolito

Manufactured in the United States of America

10 9 8 7 6 5 4 3 2

The Library of Congress has cataloged the hardcover edition as follows:

Names: Blatt, Ben.
Title: Nabokov's favorite word is mauve : what the numbers reveal about the classics,
bestsellers, and our own writing / Ben Blatt.
Description: New York : Simon & Schuster, 2017. | Includes bibliographical references.
Identifiers: LCCN 2016059215 (print) | LCCN 2017019277 (ebook) | ISBN 9781501105401
(Ebook) | ISBN 9781501105388 (hardcover) | ISBN 9781501105395 (paperback)
Subjects: LCSH: Authorship—Miscellanea. | Authorship—Technique. | Books—
Statistics. | Canon (Literature) | Language and languages—Word frequency.
Classification: LCC PN165 (ebook) | LCC PN165 .B55 2017 (print) | DDC 809—dc23
LC record available at https://lccn.loc.gov/2016059215

ISBN 978-1-5011-0538-8
ISBN 978-1-5011-0539-5 (pbk)
ISBN 978-1-5011-0540-1 (ebook)

For my mother, Faith Minard.
And for my friends at 44 Bow Street.

Contents

Nabokov's

Favorite

Word Is

Mauve

Introduction

Alexander Hamilton, James Madison, or John Jay?

For more than 150 years, historians argued over the authorship of 12 essays in *The Federalist Papers*, founding documents in the American march toward democracy. Though the essays are world-famous hallmarks in the lexicon of American history, the specific authors of each one remained unknown. The question of which Founding Father penned the essays had sparked such endless debate that it had devolved into a popular parlor game among historians. Just who exactly wrote the stirring arguments upon which our governing structure was based?

The answer was hidden in the words themselves—but to find them, scholars needed not a close reading, but a close counting. They needed to look only at the numbers.

The mystery began in late 1787, when a series of essays advocating the ratification of the Constitution was published in New York newspapers under the pen name "Publius." Shielding the true identities of the authors with the patriotic nom de plume was a somewhat farcical endeavor. In fact, of the near 4 million people living in the United States in 1787, all but three could be eliminated from contention.

It was an open secret that Hamilton, Madison, and Jay were the authors, but none of the three wanted to step forward and admit to writing any particular essays. Each had political ambitions, later rising to the ranks of Secretary of the Treasury, President, and Chief Justice of the Supreme Court, respectively, so they weren't without good reason. But their excess of caution left the mystery of authorship intact, titillating history professors and armchair enthusiasts alike for many years to come.

You might think that the scholars and astute politicos of the day would have been able to determine the authorship on their own. There were only three potential candidates, after all, each with his own political slant and style of communication. It would have been the equivalent of an anonymous editorial in the *New York Times*, penned by Barack Obama, Hillary Clinton, or Bernie Sanders. Or an unsigned manifesto by George W. Bush, John McCain, or Donald Trump. All might be coming from the same side, but they were certainly not all identical.

In 1804, a solution finally seemed to emerge. Hamilton wrote a letter to his friend Egbert Benson listing the author of each essay. Hamilton was preparing to duel Aaron Burr. He sensed both the historical significance of *The Federalist Papers* and the chances of his survival. He decided not to let his knowledge of the authorship die with him.

This should have been the end of the mystery. A nation of curious observers had no reason to doubt Hamilton's firsthand knowledge. Yet 13 years later, soon after ending his second term as President, Madison put out his own list of authorship—one that differed from Hamilton's. Twelve of the essays that Hamilton claimed to have written were also claimed by Madison.

This reopened the debate with a new fervor, fueling spats among historians for more than a century. In 1892, future senator

Henry Cabot Lodge wrote on the topic siding with Hamilton, while noted historian E. G. Bourne went with Madison.

Most historians tried to tease out the authors based on the political ideology presented in each essay. Would Madison really have argued for a central bank in those certain terms? Would Hamilton have supported limits on Congress so freely? Or maybe that's something John Jay would have written?

It wasn't until 1963, two centuries later, that the mystery was at long last solved. The definitive answer came from respected professors Frederick Mosteller of Harvard University and David Wallace of the University of Chicago. However, unlike the many professors who had attempted to solve the question before them, Mosteller and Wallace were not historians. They were not known for their scholarly work on early America. They had never published a paper on historical figures at all. Mosteller and Wallace were statisticians.

One of Mosteller's most noteworthy papers dealt with the World Series and whether or not seven games was enough to statistically find the best baseball team. Just a few years prior to looking into the authorship problem, Wallace had published a paper named "Bounds on Normal Approximations to Student's and the Chi-Square Distributions," which probably sounds as close to nonsense to you as the thought of probability functions solving a historical mystery sounded to history professors in 1963.

Mosteller and Wallace's methodology for ending the authorship debate had nothing to do with politics or ideologies. Instead, they were two of the first statisticians to leverage word frequency and probability.

Their process was in some ways complex, featuring equations with factorials, exponents, summations, logarithms, and t-distributions. But the heart of their methods was strikingly simple:

- Count the frequency of common words in essays that we know either Hamilton or Madison wrote.

- Count the frequency of those same words in essays where the author is unknown.
- Compare these frequencies to determine the author of the disputed essays.

Even before any of the fancy probabilistic equations come into play, the results of the statisticians' approach seem wonderfully obvious in retrospect. In *The Federalist Papers*, Madison used the word *whilst* in over half the essays in which his authorship had been confirmed—but he never once used the word *while*. Hamilton, meanwhile, used the word *while* in about one-third of his essays but never once used *whilst*.

Mosteller and Wallace did not rely on a single word for their analysis, however. That would not have been statistically sound. Instead, they systematically chose dozens of basic words and then found the frequency of each in the disputed essays. Many words, entirely nonpolitical in meaning, turned out to have drastically different usage rates between the two authors. For example, Madison used *also* twice as often as Hamilton, while Hamilton used *according* much more frequently than Madison.

Mosteller and Wallace had falsifiability on their side. They could show that by using the same methods on papers where the author *was* known, they could determine the authorship with perfect results. Of the 12 disputed essays, Mosteller and Wallace concluded that James Madison was the actual author of all 12.

In the written summary of their results the two mathematicians proceeded with caution, perhaps out of fear of angering historians who had been scratching their heads for generations. The numbers presented in their experiment showed a different story; the two had complete confidence in the method. It was flawless in all the test cases where authorship was known, and its results were consistent in the essays with unknown authorship. Hamilton's claim of authorship was wrong.

Today, after countless more studies of the papers in both statistical and nonstatistical manners, Mosteller and Wallace's findings—that Madison was the author—have become the consensus among statisticians and historians alike. Mosteller and Wallace were ahead of their time. Their study, though it involved some formulaic complexity, relied essentially on counting words. With today's computers, word counts and frequencies are trivial pursuits. In 1963, this was not the case.

Word counts were done by hand; to find the number of times the word *upon* appeared in each of the essays, for example, they tallied the usage page by page. To understand what Mosteller and Wallace went through (or at least what their research assistants went through), I printed out a complete collection of *The Federalist Papers* and set out to count the number of times *upon* appeared. After 30 minutes I was only one-eighth of the way through—about 40 pages—and had counted 37 instances of the word *upon*. It wasn't long before my eyes were pounding and my brain went numb. "Where's Upon?" was like a devilish version of "Where's Waldo?"

I gave up on pretending I was in 1963. Instead I did some counting only possible with twenty-first-century technology: I went to Google, searched "Federalist Papers Complete Text File," downloaded a link from the first result, and opened the file in Microsoft Word. After a grand total of two minutes, a "Find All" on *upon* turned up 46 occurrences of the word in the section I had covered. Not only was the computerized method 28 minutes faster, it was far more accurate than my weary eyes could be.

Even more staggering: Though the amount of time needed for a person to scan *The Federalist Papers* in full for an additional word would hover around four hours, scanning via computer for all words would take a negligible amount of time. Doing a similar analysis on the complete works of Shakespeare, the Bible, *Moby Dick*, or even the corpus of English literature would have been unfathomable to Mosteller and Wallace. Today, using computers to count the instances of a single word in a large text is a task mastered by most teenagers.

In the fifty years since Mosteller and Wallace published their study, the field of computer-aided text processing has grown rapidly. Google uses text analysis both in its search results and in deciding what ads to show you. Researchers have tried to use text analysis to determine what makes a tweet go viral, while media outlets often run similar versions of the same headline with slight tweaks in wording to maximize page views. But the uses thought up so far by tech companies are only one possible route.

Mosteller and Wallace used statistics to investigate a singular question of authorship. The success of their experiment was more profound. Writers have distinct styles that are both consistent and predictable. As it turns out, it's not just eighteenth-century politicians that leave a stylistic fingerprint. Authors of all books, whether they be popular and renowned or obscure and reviled, repeat their words and structure over decades of writing.

The question Mosteller and Wallace asked and answered was limited in its scope, but text analysis can answer a huge range of questions that have intrigued curious writers and readers for generations. Did Ernest Hemingway actually use fewer adverbs than other writers? How does reading level affect the popularity of a book? Do men and women write differently? Do writers follow their own advice, and is that advice any good? What, besides superficial spellings, distinguishes American and British novelists? From Vladimir Nabokov to E L James, what are our favorite authors' favorite words?

While there has been a slowly growing movement in academia to investigate the writing patterns of successful authors, there are still enormous questions that have yet to be explored. And for everyone from the casual reader to the literature major to the aspiring writer, these questions are both fascinating and useful. You probably don't care about the Poisson distribution or the parsing programs used to decipher parts of speech, but you probably do want to know how your favorite author writes—and what that might mean about you as a reader.

The analytical approach to writing can be amusing and informative and often downright funny. Moreover, it can teach us about the writers we read every day and the words we use in our own writing. That's what we'll delve into in this book, devoting each chapter to a new literary experiment.

The research won't be painfully complex. It doesn't need to be, and shouldn't be, in order to be worthwhile. Many obvious and intriguing questions about classic literature or the modern best-seller can be viewed through a statistical lens but just haven't been framed that way yet. This book is about tackling these simple yet unique questions in a new way. It's a book about words that is, paradoxically, written with numbers.

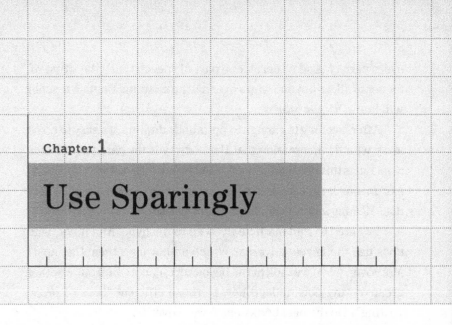

Chapter 1

Use Sparingly

> *The road to hell is paved with adverbs.*
> —STEPHEN KING

In literary lore, one of the best stories of all time is a mere six words. "For sale: baby shoes, never worn." It's the ultimate example of *less is more*, and you'll often find it attributed to Ernest Hemingway.

It's unclear whether it was in fact Hemingway who penned these words—the story of its creation did not appear until 1991—but it's natural that writers and readers would want to attribute the story to the Nobel winner. He's known for his economical prose, and the shortest-of-short stories is, at the very least, emblematic of his style.

Hemingway's simple style was an intentional choice. He once wrote in a letter to his editor, "It wasn't by accident that the Gettysburg address was so short. The laws of prose writing are as immutable as those of flight, of mathematics, of physics." He believed that writing should be cut down to the bare essentials and that extra words end up hurting the final product.

Ernest Hemingway is far from alone in this belief. The same idea is raised in high-school classrooms and writing guides of every variety. And if there's one part of speech that's the worst offender of all, as anyone who's ever had an exacting English teacher will know, it's *the adverb*.

After listening to enough experts and admirers, it's easy to come away with the impression that Hemingway is the paragon of concision. But is this because he succeeded where others were tempted by extraneous language, or is he coasting on reputation alone? Where does Hemingway rank, for instance, in his use of the dreaded adverb?

I wanted to find out if he lived up to the hype. And if not, who does use the fewest adverbs? Which author uses them the most? Moreover, when we look at the big picture, can we find out whether great writing does indeed hew to those efficient "laws of prose writing"? Do the best books use fewer adverbs?

I looked around and found that no one had ever attempted to determine the numbers behind these questions. So I sought to find some answers—and I started by analyzing the almost one million words in Hemingway's ten published novels.

If Hemingway believes that the "laws of prose writing are as immutable as those of flight, of mathematics, of physics," then I'd like to think he'd find this mathematical analysis equal parts illuminating and outlandish.

It's outlandish at first glance because of the way we study writing. Many of us have spent days in middle school, high school, and college English classrooms dissecting a single striking excerpt from a Hemingway novel. If you want to study a great author's writing, their most remembered passages are often the best place to start. Looking at a spreadsheet of adverb frequencies, on the other hand, won't teach you much in the way of writing a novel like Hemingway.

But from a statistician's point of view, it's just as outlandish to focus on a small sample and never look at the whole picture. When you study the population of the United States, you wouldn't look at just the population of a small town in New Hampshire for an understanding of the entire country, no matter how emblematic of the American spirit it may seem. If you want to know how Hemingway writes, you also need to understand the words he chooses that have not been put under the microscope. By looking at adverb rates throughout all his books, we can get a better sense of how he used language.

So instead of digging through snippets of Hemingway's text and debating specific spots where he chose to use or shirk adverbs, I used a set of functions called Natural Language Toolkit to count the number of adverbs in all of his novels. The toolkit relies on specific words and the relationships between them to tag words with a part of speech. For example, here's how it processes the previous sentence:

The toolkit relies on specific words and the relationships

—determiner —noun —verb —preposition —adjective —noun —conjunction —determiner —noun

between them to tag words with a part of speech.

—preposition —personal pronoun —to (infinitive) —verb —noun —determiner —noun —preposition —noun

It's not 100% perfect—so all the numbers below should be seen with that wrinkle in mind—but it's been trained on millions of human-analyzed texts and fares as well as any person could be expected to do. It's considered the gold standard in sussing out if a word is an adjective, adverb, personal pronoun, or any other part of speech.

So what do we find when we apply the toolkit to Hemingway's complete works?

In all of Hemingway's novels, he wrote just over 865,000 words and used 50,200 adverbs, putting his adverb use at about 5.8% of all words. On average, for every 17 words Hemingway wrote, one of them was an adverb.

This number without context has no meaning. Is 5.8% a lot or a little? Stephen King, an outspoken critic of adverbs, has a usage rate of 5.5%.

It turns out that by this standard King and Hemingway are not leaps and bounds ahead of other writers. Looking at a handful of contemporary authors who one might assume (based on stereotype alone) would use an abundance of adverbs, we see that King and Hemingway are not anomalous. E L James, author of the erotica novel *Fifty Shades of Grey*, used adverbs at a rate of 4.8%. Stephenie Meyer, whom King has called "not very good," used adverbs at a rate of 5.7% in her Twilight books, putting her right between the horror master and the legendary Hemingway.

Expanding our search, Hemingway used more adverbs than authors John Steinbeck and Kurt Vonnegut. He used more adverbs than children's authors Roald Dahl and R.L. Stine. And, yes, the master of simple prose used more adverbs than Stephenie Meyer and E L James.

All the sentences above are true—but they also need a giant asterisk next to them and a full explanation. Because the answer is not as simple as the numbers above first suggest.

Those tallies are counts of total adverb usage. An adverb is any word that modifies a verb, adjective, or another adverb—and no adverbs were excluded or excused. But when Stephen King says, "The adverb is not your friend," he's not talking about any word that modifies a verb, adjective, or another adverb. In the sentence "The adverb is not your friend," the word *not* is an adverb. But *not* is not King's issue. Nobody reads "For sale: baby shoes, never worn" and thinks *never* is an adverb that should have been nixed.

When King rails against adverbs in his book *On Writing*, he describes them as "the ones that usually end in -ly." From a statistical standpoint his "usually" isn't quite true (depending on the author, around 10 to 30% of all adverbs are ones that end in -ly) but it is true that the adverbs ending in -ly are the ones that tend to stick out.

Chuck Palahniuk, best known as the author of *Fight Club*, has written against -ly adverbs as well. When discussing the importance of minimalism in his book *Stranger than Fiction*, Palahniuk writes, "No silly adverbs like 'sleepily,' 'irritably,' 'sadly,' please." His general argument is that writing should allow us to know when a character is sleepy, or irritable or sad, by using a broader set of clues rather than a single word. Using -ly adverbs goes too far, telling the reader what they should think instead of setting up the scene so that the meaning becomes clear in context.

By narrowing our search to just -ly adverbs, we can cut to the heart of the debate. And when we do, the picture flips. For every 10,000 words E L James writes, 155 are -ly adverbs. For Meyer the count is 134, while King averages 105. And Hemingway, living up to his reputation, comes in at a scant 80.

Below, for the sake of comparison, is a breakdown of adverb use among 15 different authors.

Number of -ly Adverbs per 10,000 Words

AUTHOR	BOOKS	
Ernest Hemingway	10 Novels	80
Mark Twain	13 Novels	81
Amy Tan	6 Novels	83
John Steinbeck	19 Novels	93
Kurt Vonnegut	14 Novels	101
John Updike	26 Novels	102
Salman Rushdie	9 Novels	104
Stephen King	51 Novels	105
Charles Dickens	20 Novels	108
Virginia Woolf	9 Novels	116
Herman Melville	9 Novels	126
Jane Austen	6 Novels	128
Stephenie Meyer	4 Twilight Books	134
J. K. Rowling	7 Harry Potter Books	140
E L James	3 Fifty Shades Books	155

Looking at this strict definition of the "bad" kind of adverb, Hemingway indeed comes out as one of the greats. As we continue to explore in this chapter, whenever I use the term *adverb*, I will be referring to this "bad" sort—the -ly adverb.

Was Hemingway Right?

The list on the previous page includes a variety of writers, from Nobel Prize winners to viral bestsellers. Hemingway may emerge as a titan of unadorned prose, just as the common perception of him would suggest. But any broader pattern is not so clear. E L James lands at the top of the scale, but greats like Melville and Austen also clock in toward the higher end. By adding more data points, would we be able to pinpoint a reliable pattern in adverb usage?

I wanted to find out whether an author's adverb rate reflects anything more than just personal style or preference. I was curious: Could Hemingway have been right about the "laws of prose"? Is there any meaningful relationship between the quality of a book and how often it uses adverbs?

To start answering these questions, it's important to note that just as different authors vary in their use of adverbs, so do different books by the same author. The rate of these -ly adverbs is rare—under 2%—even for authors who use them more than other scribes. And there is often great variation from book to book within an author's career.

For instance, looking at Hemingway's novels, we see a wide range. Several of his books have adverb rates much lower than most authors ever come close to approaching, while other books bounce around the average usage rate of other authors. *True at First Light*, a novel about Hemingway's experiences in Africa, is his novel with the highest adverb usage—and it's one released thirty years after his death.

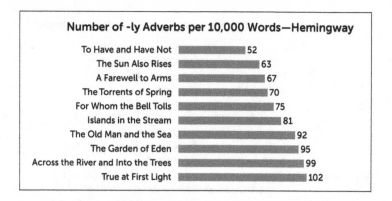

Number of -ly Adverbs per 10,000 Words—Hemingway

To Have and Have Not	52
The Sun Also Rises	63
A Farewell to Arms	67
The Torrents of Spring	70
For Whom the Bell Tolls	75
Islands in the Stream	81
The Old Man and the Sea	92
The Garden of Eden	95
Across the River and Into the Trees	99
True at First Light	102

True at First Light was received by critics with negative reviews. It was unfinished at the time of Hemingway's death and edited into shape by his son. Some saw its publication as an unnecessary addition to the canon. Is it a coincidence that it is also his work with the most adverbs?

It's of course a poor criterion to judge a book on nothing but its adverb rate, but looking at Hemingway's complete works, we see that most of his classics are also the texts in which he uses the fewest adverbs. *The Sun Also Rises*, *A Farewell to Arms*, and *For Whom the Bell Tolls* have some of the lowest rates—and are considered to be among Hemingway's best. *The Old Man and the Sea*, which won the author the Pulitzer Prize and is often named Hemingway's best work, is the exception.

The two American authors to win the Nobel Prize in literature within a decade of Hemingway are William Faulkner and John Steinbeck, and we can pick apart their stats as well.

For Steinbeck, the rate of adverb usage again matches up well to perceptions of his work. *The Grapes of Wrath*, perhaps his most popular work, places third on the list. *Of Mice and Men* and *East of Eden* also land toward the low end.

For Faulkner, the pattern is again present. His most celebrated work, *The Sound and the Fury*, ranks second with a low 42 ad-

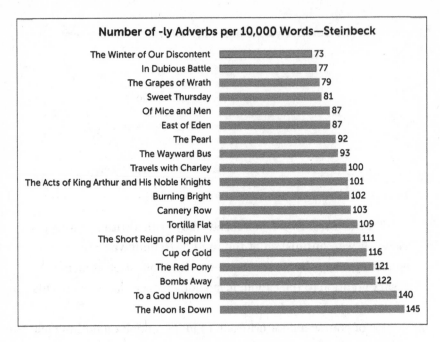

Number of -ly Adverbs per 10,000 Words—Steinbeck

The Winter of Our Discontent	73
In Dubious Battle	77
The Grapes of Wrath	79
Sweet Thursday	81
Of Mice and Men	87
East of Eden	87
The Pearl	92
The Wayward Bus	93
Travels with Charley	100
The Acts of King Arthur and His Noble Knights	101
Burning Bright	102
Cannery Row	103
Tortilla Flat	109
The Short Reign of Pippin IV	111
Cup of Gold	116
The Red Pony	121
Bombs Away	122
To a God Unknown	140
The Moon Is Down	145

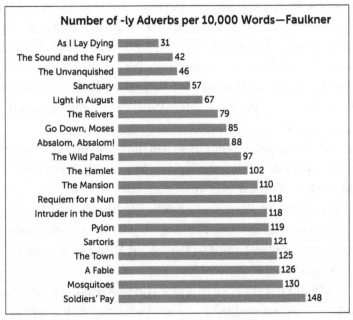

Number of -ly Adverbs per 10,000 Words—Faulkner

As I Lay Dying	31
The Sound and the Fury	42
The Unvanquished	46
Sanctuary	57
Light in August	67
The Reivers	79
Go Down, Moses	85
Absalom, Absalom!	88
The Wild Palms	97
The Hamlet	102
The Mansion	110
Requiem for a Nun	118
Intruder in the Dust	118
Pylon	119
Sartoris	121
The Town	125
A Fable	126
Mosquitoes	130
Soldiers' Pay	148

verbs per 10,000. *As I Lay Dying* and *Light in August* also come in at the top, while *Absalom, Absalom!* is just under his average as well.

But this is just three authors. How far does the pattern go? If we expand outward, do the best books by the best authors use fewer adverbs on average?

The authors selected for the tables on the previous pages, and the books highlighted, are some of my own favorites, chosen based on my own preferences. To test whether adverb rates have any correlation with writing quality, I would need a larger set of books and writers—and I would need them to be considered "great" by a consensus of readers.

To build a new sample, I turned to four different lists of the best twentieth-century literature: the *Library Journal* list, the Koen Book Distributor's list, the Modern Library List, and the Radcliffe Publishing Course list. All four lists rank at least 100 works of fiction in English literature. These four lists were also used by Stanford librarian Brian Kunde in his attempt to quantify the best book of the twentieth century (by his scoring it's *The Great Gatsby*). For my purposes, if a book was included on at least two of the four lists I deemed it a consensus "great" book. I then selected the authors who had at least two of these, so I would be able to compare their "great" books to the "non-great." (Of course, this method leaves out a lot of excellent authors, but I needed something approaching "objective" and I needed authors with multiple works.)

The result is 15 consensus "great" authors with a total bibliography* of 167 fiction books, of which 37 books were considered "great" by virtue of being on multiple top-100 lists. Here, I've listed those 37 "great books."

---

* Some of Sinclair Lewis's novels could not be found in digital form and were excluded.

The Consensus "Great" Books

Willa Cather
Death Comes for the Archbishop
My Ántonia

E. M. Forster
A Passage to India
A Room with a View
Howards End

Toni Morrison
Beloved
Song of Solomon

Joseph Conrad
Heart of Darkness
Lord Jim

Ernest Hemingway
A Farewell to Arms
For Whom the Bell Tolls
The Old Man and the Sea
The Sun Also Rises

George Orwell
Animal Farm
1984

Theodore Dreiser
An American Tragedy
Sister Carrie

James Joyce
A Portrait of the Artist as a Young Man
Ulysses
Finnegans Wake

Ayn Rand
Atlas Shrugged
The Fountainhead

William Faulkner
As I Lay Dying
Light in August
The Sound and the Fury

D. H. Lawrence
Lady Chatterley's Lover
Sons and Lovers
Women in Love

John Steinbeck
Of Mice and Men
The Grapes of Wrath

F. Scott Fitzgerald
Tender Is the Night
The Great Gatsby

Sinclair Lewis
Babbitt
Main Street

Edith Wharton
Ethan Frome
The Age of Innocence
The House of Mirth

The test I conducted was simple but revealing, aimed at parsing whether there is a noticeable difference between the best books and the rest. I combined all 167 books written by the consensus "great" authors, and I broke them down into groupings of 50 adverbs per 10,000 words. I then looked at how many books in each group were selected as "great" versus "nongreat." The graph on the following page charts the results. Books with 0–49 adverbs per 10,000 words were considered great by critics 67% of the time. Those in the 50–100 range were selected as great 29% of the time. On the far end, those with 150-plus received the honor just 16% of the time.

The downward slope of the graph backs up the advice of Hemingway, King, and countless other writers. While far from absolute, there's a clear trend as adverb use increases. The best books—the

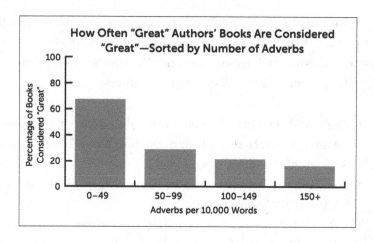

How Often "Great" Authors' Books Are Considered "Great"—Sorted by Number of Adverbs

greats of the greats—*do* use a lower rate of -ly adverbs. On the other hand, an overuse of adverbs has resulted in "great" books at a far scarcer frequency.

Author by Author, Adverb by Adverb

Looking at the best books all in one graph gives us a compelling picture, but it is still just part of the picture. The greats may tend to use fewer adverbs, but it's also clear that a book or author doesn't *need* to adopt the same low rate to be great. Consider Sinclair Lewis, a Nobel Prize winner and one of our consensus "great" authors, who writes at an average adverb rate of 142 per 10,000 words. That's *a lot*—75% more than Hemingway averaged.

At the very least, Lewis is an outlier—perhaps even an argument against any general trend. But when we dig into his work, we find a pattern that seems to apply even to the previous section's outliers. What's interesting about Lewis's work is that his two best books—*Main Street* and *Babbitt*, his two consensus "great" books—use fewer adverbs than any of his other novels. In other words, even though Lewis uses a very high rate of adverbs, his most popular writing is his most concise.

With Hemingway, Steinbeck, and Faulkner, we have already seen that their masterpieces also tend to be their books with fewer adverbs. And, looking beyond this trio, we find plenty of similar examples throughout the great authors:

- *The Great Gatsby* is F. Scott Fitzgerald's book with the lowest adverb rate. His second most popular novel, *Tender Is the Night*, is his book with the second lowest rate.
- Toni Morrison's most acclaimed novel, *Beloved*, is tied as her book with the fewest adverbs.
- *A Tale of Two Cities* and *Great Expectations* beat out the other 13 Charles Dickens novels to have the lowest and second lowest adverb rates.
- Kurt Vonnegut wrote 14 novels, and his three most acclaimed are *Cat's Cradle*, *Slaughterhouse-Five*, and *Breakfast of Champions*. They rank first, second, and third in least adverb usage out of all his works.
- John Updike authored 26 novels. The four novels with the smallest adverb rate were all four books in his Pulitzer Prize–winning Rabbit tetralogy.

The string of examples goes on, but there are also notable exceptions. D. H. Lawrence, for instance, wrote two "great books," in *Lady Chatterley's Lover* and *Women in Love*, that use *more* adverbs than any of his other works. If we continue to search, we can find anecdotal evidence for either side.

The Sinclair Lewis question, then, was the next big trend I wanted to investigate. Lewis, it seemed, might reveal an even broader truth about how authors use and abuse their adverbs: Regardless of an author's fondness for adverbs—whether their natural rate skews high or low—are they at their *most successful* when they're most concise?

To test this, I would need to go beyond the simple distinction between "great" and "non-great" books. I would need to be able to compare all novels within an author's bibliography on a sliding scale, measuring how good the "great" books are and how bad the "non-great" are. Doing so, it would be possible to chart out whether—book by book, within an individual author's career—there's a broader correlation between adverb use and writing quality.

How, though, can you compare any two books in an objective manner? When neither appears on any critic's "best of" list, how do we know with any reliability whether Steinbeck's *The Pearl* can be considered better than his novel *To a God Unknown*?

The solution I settled upon was to dig into the ratings found on book reviewing sites like Amazon or Goodreads. Goodreads .com is a website where people go to rate, discuss, and catalog books. Here, a popular book can have more than one million ratings, a quantity much larger than the same book would receive on Amazon. Because of its size, we'll explore Goodreads ratings.

In particular, I've chosen to focus on how *many* Goodreads ratings a book has. It's not a perfect metric, but it gives a relative sense of the reception and popularity of a book. In fact, it's a better metric than the book's average numerical rating—the average number of stars it receives from reviewers.* Using Goodreads data, we're no longer stuck with the binary of a book being "great" or "not great." We can instead get a sense of a book's popularity on a spectrum, which gives us a much fuller picture of its quality and status.

We can then go back to our Steinbeck and Faulkner graphs and bulk them up with more depth. The books farthest to the left have the most ratings while the books to the right have the least. The books closest to the top have the fewest adverbs while those

---

* It's more wonky than we can get into here, but for a detailed explanation of why average rating falls short, head to the Notes section on p. 251.

at the bottom have the most. For Steinbeck, *The Grapes of Wrath* has many ratings and few adverbs so it's in the upper left. *Bombs Away* has few ratings but lots of adverbs, so it's off in the lower right-hand corner.

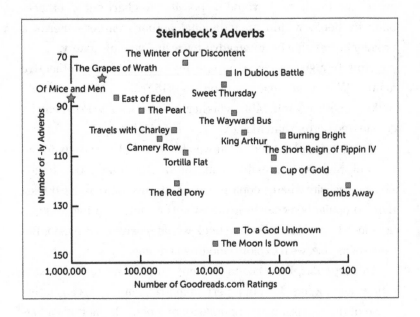

If the correlation were perfect, the books would form a pattern always trending down and to the right when looking at the chart. It's not exact, but the correlation exists.

The number of ratings in the graph above is displayed using a logarithmic scale, meaning a book with 100,000 ratings will be just as far off from a book with 10,000 ratings as a book with 10,000 ratings would be from one with 1,000. Without the logarithmic scale, the books would be too far apart to make any sense of it. If you're having trouble wrapping your head around the logarithmic scale, the Faulkner graph shows the same trend but instead breaks the data down into rankings—the book with the most Goodreads ratings is considered number one on the horizontal axis, then the next highest is number two, and so on. If the correlation were perfect, every book would fall on the dashed line.

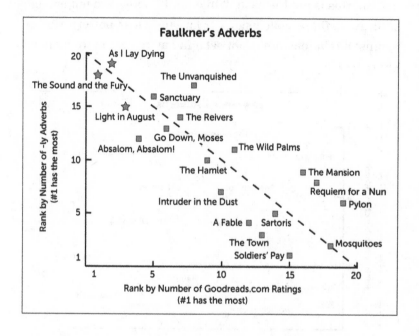

Even if you were to look at just the "non-great" books—excluding *As I Lay Dying*, *The Sound and the Fury*, and *Light in August*—the relationship holds in stunning fashion. The empty lower left-hand quadrant shows the complete lack of Faulkner books that have both high -ly adverb rates and a noteworthy reputation.

This amazing trend does not hold for every author. If there was one author poised to buck the trend, based on the complete listings of adverb use above, it would be D. H. Lawrence. The Englishman was unique among our 15 authors in that his book with the *most* adverb usage was considered "great." And his book with the second-most adverbs was considered "great" as well. You might then expect that the rest of his work would follow that same trend, forming a line from the bottom left corner to the top right (the opposite of Faulkner's).

But this is not the case. While his 12 books are not enough to draw definite conclusions from, any clear pattern (for or against a relationship) is not seen in Lawrence's more jumbled chart.

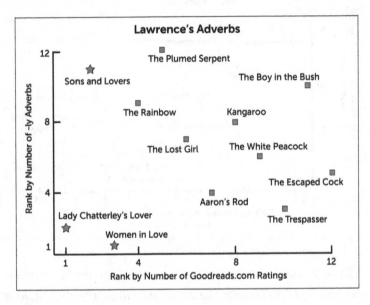

Though Faulkner's graph may appear to have an undeniable pattern at first glance, the brain can sometimes play tricks searching for patterns in images, so it's always better to test for significance to see what the raw numbers say. In Faulkner's case the numbers back up the eye test. There is a correlation between adverb use and his book ratings (going back to log scale) when tested. Lawrence's graph, on the other hand, has no pattern to speak of. But which of these graphs is the norm? Is Lawrence the outlier, or are Faulkner and Steinbeck anomalies?

The bigger picture comes together when we combine all the authors into a single graph and normalize their adverb usage and

Goodreads popularity.* With this approach we can look at all our great authors and all their books, and we can ask whether there is a correlation between adverb use and book quality within each author's career.

In the full sample of 167 books, we find that there is indeed a correlation. The pattern isn't perfect, but the connection is striking—and it goes well beyond the variation we could expect due to chance alone. Up and down the sample, we find that authors' books with the least adverbs have been their most popular, and their books with more adverbs have tended to earn lower ratings.

The chart below illustrates the general pattern, as well as highlighting the large number of outliers. The top left-hand corner shows books that ranked in the top half of an author's most popular works and are in the bottom half of adverb usage. Fifty books fall in this range. In contrast, just thirty-one books are among the most popular half but also high in adverbs. The hits are concise, while the wordier novels are often forgotten.

High Goodreads Ranking Low in Adverbs 50	Low Goodreads Ranking Low in Adverbs 30
High Goodreads Ranking High in Adverbs 31	Low Goodreads Ranking High in Adverbs 50

Note: The numbers do not sum to 167 books because some books were at the median—meaning they could not be categorized.

---

* Unlike the previous section's aggregate graph, where books from different authors were combined unadjusted, the normalization here allows us to compare authors of different levels of popularity and adverb rate without any outliers skewing the combined chart. The authors are treated as if they have equal rates and popularities, so that we can concentrate on the trends *within* each author's work.

Pros vs. Amateurs

We've now seen that adverb use does play a role in the work of the canon's best authors and their greatest works. At the pinnacle of the literary world, the standout books indeed rely on fewer adverbs. And even within each author's own works, the books that use the least adverbs have been the most successful.

But there was one more question on my mind: What about the rest of us?

Before I'd be satisfied, I wanted to find out how great writers compare to the average writer when it comes to adverb use or abuse. Do Hemingway's "laws of prose" apply across the whole of the literary universe, from the award winners and bestsellers to the amateurs? I set up one final showdown to find out.

I downloaded more than 9,000 novel-length fan-fiction stories (of 60,000-plus words) from fanfiction.net. This would be my "amateur" group, consisting of all stories written between 2010 and 2014 in the 25 most popular book universes (ranging from Harry Potter to Twilight to *Phantom of the Opera* to Janet Evanovich's books). People writing stories this long are committed to their work, and many of them are strong writers. But on average, they're not at the level of the bestsellers or the award winners of the literary world. So I compared the fan-fiction sample to all of the books that have ranked number one on the *New York Times* bestseller list since 2000, and also to the 100 most recent winners of major literary awards.*

When set side by side, the difference is clear. The median fan-fiction author used 154 -ly adverbs per 10,000 words, which is much higher than either of the professional samples. The 300-plus megahits in the bestseller category averaged just 115 -ly adverbs

---

* The selections for these award-winning books is described in detail in Chapter 2.

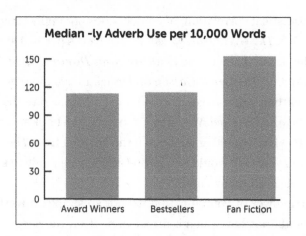

Median -ly Adverb Use per 10,000 Words

per 10,000 words. And the 100 award winners have a median of 114 -ly adverbs. It's not an apples-to-apples comparison, but the novels that sell well in bookstores come in with 25% fewer adverbs than the average novel that amateur writers post online. Less than 12% of all number one bestsellers had more than 154 adverbs, even though half of all fan fiction does.

The results of this chapter are one half common sense and one half mind-blowing.

Most writers and teachers will tell you that adverbs are bad. This is not a controversial stance to take. In many ways, the statistics presented above are just a confirmation of what we already knew.

But the fact that their use is somehow correlated with quality on a measurable level—even when just the best writers are being examined—is still shocking. It might not be a surprise that some beginner writers use adverbs as a crutch more often than professional writers, and that these traits may sometimes be noticeable. But even when looking at the life's work of the best writers, the effect is present.

A statistical correlation, of course, does not imply causation. Fitzgerald's *The Great Gatsby* used 128 adverbs per 10,000 words while his lesser-known *The Beautiful and Damned* used 176. If you picked up *The Great Gatsby* and stuck in 200 more adverbs, a bit less than one a page, it would have a higher rate than *The Beautiful and Damned*. Would this version of the book still be celebrated? What if you trimmed down the adverbs from *The Beautiful and Damned*? Would Leonardo DiCaprio be ready to suit up for the role of Anthony Patch?

The answer of course is that it's not so simple. Adverb rate alone could not have such a direct impact on the success of a book. There are thousands and thousands of other aspects of writing in play. The Hemingway adverb stereotype may be true, but there are notable counterexamples—authors who have written successful books when increasing their adverb usage. Nabokov's *Lolita*, for instance, has more adverbs than any of his other eight English novels.

One possible explanation for the overall trend we're seeing is that adverbs are an indicator of a writer's focus. An author writing with the clarity needed to describe vivid scenes and actions without adverbs, taking the time to whittle away the unnecessary words, might also be spending more time and effort making the rest of the text as perfect as possible. Or if one has a good editor, these words may be weeded out.

The "focus" hypothesis finds some support from the true master of writing without adverbs. And it's not Hemingway.

The numbers revealed an overlooked champion. Combing through a large number of authors, there was but one author on the list of "greats" who outdid Hemingway: Toni Morrison. She may be a Nobel and Pulitzer Prize winner just like Hemingway, but her place at the height of concise writing isn't often cited in English classrooms. Her adverb rate of 76 edges out Hemingway's 80, and puts her well ahead of others like Steinbeck, Rushdie, Salinger, and Wharton.

Morrison has said in multiple interviews that she doesn't use adverbs. Why? Because when she's writing at her best, she can do without: "I never say 'She says softly,'" Morrison tells us. "If it's not already *soft*, you know, I have to leave a lot of space around it so a reader can hear that it's soft."

There you have it. And while I have no hard evidence that the logic of adverb usage carries over to wacky statistics-based prose, I went through the text of this chapter to search for -ly adverbs after writing 5,000 words on how awful they were. I found that in most cases they were unneeded. They often blunted the impact of my sentences. I deleted all -ly adverbs that were not used when quoting or citing others.

As a result, if you excuse the ones in quotes, you will find no -ly adverbs in this chapter. This makes for a usage rate of 0 per 10,000 that would rank this text ahead of (or tied) with all other texts ever written. Does that make this chapter, regardless of content, a step above average? Here we've found the limits of our statistics. But when trying to write standout prose, it can't hurt to ~~deliberately~~ avoid the troublesome part of speech.

Chapter 2

He Wrote, She Wrote

> *It is fatal for anyone who writes to think of their sex.*
> —VIRGINIA WOOLF, *A ROOM OF ONE'S OWN*

Let's say we have two Facebook statuses. One is written by a woman, one by a man. You've been offered five dollars if you can guess which post is which; but you'll only be given a short selection of words from each post. Given the samples below, would you be able to win that five dollars?

Selection 1: *shit, league, shave*
Selection 2: *shopping, boyfriend, <3*

You'd feel pretty confident in your ability to make a guess, right?

Now, what if you were asked to perform the same task, but shown these triplets instead?

Selection 3: *actually, everything, their*
Selection 4: *above, something, the*

There are fewer clues. But what if I told you that there's a clear best guess?

*　　*　　*

For generations researchers have been studying the ways in which men and women differ in their writing—often with little concrete evidence to show for it. In recent years, however, computer scientists have been able to comb through overflowing social media data to be able to pinpoint small differences. It's not just an academic exercise, either. The prize in this scenario is not five dollars, but billions in targeted ads. Some of the results of this research have been all too cliché and stereotypical (*shopping* skews female; *league* skews male). But some seem altogether perplexing: The innocuous words in selections 3 and 4 do indeed tend to vary between genders, and researchers have harnessed them to make shockingly accurate predictions.

I wanted to use the same methods to look at literature rather than tweets and Facebook posts. But before we make that leap, let's explore what the two examples above are looking at when they classify writing as "male" or "female."

The words in the first two selections—*shit, shave, league, shopping, boyfriend,* <3—all come from a paper published by researchers at the University of Pennsylvania that churned through millions of Facebook statuses in search of the select few words that are *most indicative of gender.* (As you probably guessed, Selection 1 contains the most "male" words, and Selection 2 contains the most "female" ones.)

That doesn't mean that all men talk *shit* in their status updates or that all women talk about *shopping.* In fact, *shopping* is not a word that's used frequently by either gender. These "most indicative" stats measure the words that one group uses often compared to how rarely other groups use them. *Shopping*'s presence on the list, in a sense, says more about the fact that men *don't* often talk about shopping than that women do. This method of discerning gender attempts to find the starkest contrasts between what men are writing about and what women are writing about. And for that reason, the findings above and in the chart below skew toward the extreme ends of gender norms.

Below, we see the top five Facebook status words for each gender, as well as similar findings gleaned from a range of different social media.

CORPUS	WORDS WITH LOPSIDED USAGE BY MALES	WORDS WITH LOPSIDED USAGE BY FEMALES
Facebook Status	Fuck, League, Shit, Fucking, Shave	Shopping, Excited, <3, Boyfriend, Cute
Chatroom Emoticons	;)	:D
Twitter Assent or Negation Terms	Yessir, Nah, Nobody, Ain't	Okay, Yes, Yess, Yesss, Yessss, Nooo, Noooo, Cannot
Blogs	Linux, Microsoft, Gaming, Server, Software	Hubby, Husband, Adorable, Skirt, Boyfriend

The studies from which these findings originate are listed in the Notes section on page 262.

The words revealed by the second two selections—*actually, everything, their, above, something, the*—are very different at first glance. They don't fall into one gender stereotype or another; rather they're function words that everyone uses. But in a 2003 paper, computer scientists looked into gender differences in writing by examining samples from the British National Corpus (both fiction and nonfiction), and they came back with some curious results. Their biggest findings dealt with these very small words. For instance, they claimed that across all genres of writing "females use many more pronouns" (*I, yourself, their*) and males use "many more noun specifiers" (*a, this, these*).

The notion that such a general conclusion could be drawn using entirely mundane words sounds absurd. However, the paper went on to show that by using the frequencies of just a few dozen tiny words, the authors were able to create an algorithm that accurately predicted an author's gender 80% of the time when examining randomized documents. That's a huge percentage—

all based on tiny words like the six found in Selections 3 and 4. (And, for the record, you'd want to bet that Selection 3 was the one penned by a female author and that Selection 4 was written by a male.)

Both of these statistical methods rely on broad generalizations, but after reading about them in depth, I was curious. How do these ways of measuring and predicting gender hold up if we apply them to even bigger, tougher samples? And can they reveal anything interesting about the state of literature?

I decided to find out what they'd come up with when we compare men and women in classic and popular fiction. What words or books would show up as the "most male" or "most female"? And would the findings of that 2003 paper be able to predict a novelist's gender with any reliability?

To explore these questions, I've identified three samples of books that we'll come back to throughout the rest of this chapter: classics, recent bestsellers, and recent literary fiction. I will refer to them in shorthand for the rest of the text, but the exact rules I've used to derive each sample are below (and full lists can be found in the Notes section at the end of the book).

Classic Literature

I started with Stanford librarian Brian Kunde's composite list of "The Best English Language Fiction of the Twentieth Century." Kunde combined the results of many literature polls to compile the list. From this aggregate list I took the top fifty novels or story collections written by men and the top fifty written by women. These are the type of books you'd see from authors like Ernest Hemingway, Willa Cather, William Faulkner, or Toni Morrison.

Modern Popular Fiction

Starting with the end of 2014 and going backward, I found the last fifty number one *New York Times* fiction bestsellers written by men and the last fifty by women. I threw out any book that had a listed co-author. The resulting collection consists of books written by blockbuster authors like Nora Roberts, Stephen King, Jodi Picoult, and James Patterson.

Modern Literary Fiction
Starting with awards given at the end of 2014 and looking backwards, I found the last fifty novels written by men and the last fifty novels written by women that were on any of the following lists: *New York Times* Top Ten Books of the Year, Pulitzer Prize finalists, Man Booker Prize short list, National Book Award finalists, National Book Critics Circle finalists, and *Time* magazine's best books of the year. The end sample of 100 books, ranging from 2009 to 2014, includes many authors like Jennifer Egan, Jonathan Franzen, Michael Chabon, and Zadie Smith.

I started with the first method given above, looking at the words used most disproportionately by each gender. This technique tends to find the extremes and stereotypes—the *shopping* and *shave* words that one gender uses most often compared to how rarely the other gender uses them.

For example, in the Classic Literature section, the word *dress* was used 2,069 total times and at least once in 97 of the 100 books. It was used at an average rate just over once per 10,000 words. While 35 female authors used it above this rate, only seven male authors did. I used this ratio to give it a score of 83% (35:7) female, and all of the top words were determined by this ratio.

I chose this methodology, instead of a pure ratio, to control for the fact that some authors use specific words often to suit their plot. For instance, Tolkien used the word *ring* over 750 times in the three *Lord of the Rings* books (which counted as a single entity in the "classic book" sample). That's more times total than it was used in all fifty classics by women. However, *Lord of the Rings* excluded, there is no evidence that men use the word *ring* more often than women (rather in all other books in the sample, female authors use it about twice as often).

I also restricted the "most male" and "most female" words to those that were used in at least fifty books in the sample of 100, so that they weren't outliers. With these rules in mind, here are the words with the most extreme imbalance in the 100 classics I've gathered:

Most Gender-Indicative Words in Classic Literature	
MALE	FEMALE
Chief	Pillows
Rear	Lace
Civil	Curls
Bigger	Dress
Absolutely	China
Enemy	Skirt
Fellows	Curtains
King	Cups
Public	Sheets
Contact	Shrugged

Many of these words appear to be driven by the plot and scope of the books they're contained within: The male words tend to skew toward the military or governmental while the female words tend to skew toward the domestic. While fifty books by each gender is a small sample, writing turns out be consistent enough over time (or at least the subjects that writers of different genders choose to write about are consistent enough) that these words remain indicators even in modern literary and popular fiction. *Chief* is used more by men than women in popular fiction and modern literary fiction. *Pillows* is used more by women in popular fiction and modern literary fiction.

Of the 20 words above, 16 of them fall on the same side of the gender-skewed aisle in popular fiction. In literary fiction, 18 of them do. If these words were distributed in a random fashion, perhaps half would fall on the other side of the divide when we look at different samples. Even in a modest-sized sample, these words are consistently used more often by one gender than the other.

These words show the lopsidedness of certain topics, but not necessarily a difference between how the two sexes write. So next, I decided to try out the second, eerier method—that of predicting gender using simple words like *it* and *is*—to see if it had any effectiveness in novels. Based on that original 2003 paper, a computer programmer named Neal Krawetz developed a quick system to guess the gender of an author using just 51 words. Drawing from the paper, Krawetz chose very common and ordinary words. Twenty-four were used at a higher rate among male writers: *a, above, are, around, as, at, below, ever, good, in, is, it, many, now, said, some, something, the, these, this, to, well, what,* and *who*. Twenty-seven words were used at a higher rate by female writers: *actually, am, and, be, because, but, everything, has, her, hers, him, if, like, more, not, out, she, should, since, so, too, was, we, when, where, with,* and *your*. For his simplified method of guessing gender, each word was given a point value based on its relative predictive value (*these* is +8 for male, while *since* is +25 for female). Each time a word is used in the text its point value is counted until a final total is tabulated.

If I wrote, **The** *method* **is** *simple* **and** *crude*, the algorithm would give that a male-female ratio of 91% (+24 male for *the*, +18 male for *is*, +4 female for *and*) whereas **The** *method* **is not too** *complicated* would have a male-female ratio of 39% (+24 male for *the*, +18 male for *is*, +27 female for *not*, +38 female for *too*).

It might be useful to reflect for a second on just how basic this system is. There is no context or other interference involved at all. It would guess that the sentence *This sentence is written by a woman* was written by a man. This is not a nuanced method. For the results below I removed the pronouns *her, hers, him,* and *she* to make sure it was not relying on simple gender giveaways. (Later in this chapter we'll see how indicative gendered pronouns are in identifying the authors of fiction.) Without them, all that's left is a handful of seemingly gender-neutral pronouns, conjunctions, and identifiers.

The computer scientists in the 2003 paper managed to predict an author's gender at a rate of 80%. I wasn't able to match that when applying Krawetz's method to classic literature. However, the system does (perhaps remarkably) succeed, managing to perform significantly better than chance. The simple method identified the correct author gender for the 100 classic books 58 times. Of the 100 modern number one bestsellers, it was right 66 times; and for the modern literary novels, it was right 58 times. Below are the classic books that it ranks as being most likely to have been written by a male or female author. In other words, these are the "most masculine" and "most feminine" novels, according to Krawetz's method.

MOST MASCULINE CLASSIC NOVELS	MOST FEMININE CLASSIC NOVELS
A Portrait of the Artist as a Young Man—James Joyce	Ellen Foster—Kaye Gibbons
Charlotte's Web—E. B. White	Rubyfruit Jungle—Rita Mae Brown
Orlando—Virginia Woolf*	A Clockwork Orange—Anthony Burgess*
Animal Farm—George Orwell	Their Eyes Were Watching God—Zora Neale Hurston
The Shipping News—Annie Proulx*	The Catcher in the Rye—J. D. Salinger*
Winesburg, Ohio—Sherwood Anderson	Bastard Out of Carolina—Dorothy Allison
Lord of the Flies—William Golding	The Color Purple—Alice Walker
Atlas Shrugged—Ayn Rand*	Wide Sargasso Sea—Jean Rhys
Death Comes for the Archbishop—Willa Cather*	Lady Chatterley's Lover—D. H. Lawrence*
The Sun Also Rises—Ernest Hemingway	The Death of the Heart—Elizabeth Bowen

* Indicates the author is of the opposite gender from what the numbers predicted.

And here is the breakdown of recent bestsellers' most masculine and feminine titles (using the Modern Popular Fiction sample). In this sample, there is just *one* error in the method's top twenty categorizations.

MOST MASCULINE #1 BESTSELLERS	MOST FEMININE #1 BESTSELLERS
Inferno—Dan Brown	*Kiss the Dead*—Laurell K. Hamilton
The Fallen Angel—Daniel Silva	*Power Play*—Danielle Steel
The English Girl—Daniel Silva	*Hit List*—Laurell K. Hamilton
The Heist—Daniel Silva	*Until the End of Time*—Danielle Steel
Act of War—Brad Thor	*Gone Girl*—Gillian Flynn
Flash and Bones—Kathy Reichs*	*Big Little Lies*—Liane Moriarty
The First Phone Call from Heaven—Mitch Albom	*Dead Ever After*—Charlaine Harris
Kill Alex Cross—James Patterson	*New York to Dallas*—J. D. Robb
Cross My Heart—James Patterson	*Frost Burned*—Patricia Briggs
The Time Keeper—Mitch Albom	*Dead Reckoning*—Charlaine Harris

* Indicates the author is of the opposite gender from what the numbers predicted.

In the original paper, the researchers—all men, who were perhaps fearful of being derided for any politically incorrect theories—wrote that they would not speculate on the reasons for gender differences "to avoid baseless speculation with regard to interpretation of the data." They did, however, go on to point to papers, written in the 1980s and 1990s, which looked at the differences in male and female conversations. The theories proposed in these papers held that men used more "informational" language, about objects, while women used more "involved" language, about relationships. (In their words, " 'Involved' documents contain features which typically show interaction between the speaker/writer and the listener/reader, such as first and second person pronouns.")

Such generalizations hint at how these words, even if they seem context free, may also be tapping into the same extremes and gender norms that our first example latched on to. For instance, *above, around*, and *below* were all male indicative words, and these are clearly "informational words" by the standards of the researchers. However, it's unclear whether these words are used more by males because, as a general rule, males "are" more informational—or whether male authors, for all sorts of cultural and historical reasons, choose to write more stories on war and physical action that require that type of detail.

You may not be familiar with all the books in the list above, but you probably know that Dan Brown writes thrillers about global conspiracies while Danielle Steel writes romances about the rich and famous. That difference in genre could explain the effectiveness of Krawetz's seemingly neutral method. Even though the 22 books in the Alex Cross series by James Patterson and 27 books in the Anita Blake series by Laurell K. Hamilton are all thrillers at the core, Hamilton's series relies much more on Blake's romantic conflict while Patterson's series only features Cross's love life as a much smaller subplot.

There's a chicken-and-egg type dilemma that sets in when trying, as researchers have for decades, to pinpoint the mysterious stylistic DNA that differentiates male writing from female writing. Even the subtler predictive method seen here seems to be reaching back to an author's initial decision about subject matter—what and who we choose to focus on in our stories—rather than revealing something more fundamental. It may seem odd that an 80% accuracy rate or even a 60% rate in novels is achievable. But consider what the following example says about *who* authors of different genders choose to write about.

In his novel *Chance*, Joseph Conrad wrote, "Being a woman is a terribly difficult trade since it consists principally of dealings with men."

It's a line that can be read as both sympathetic and unsympathetic. On the one hand, Conrad is showing awareness of unfair inequality, showing empathy for what women often put up with at the hands of men. Yet at the same time the one-liner suffers from a conceptual block, implying that a woman's main ends are not self-contained but dependent on men. It reinforces that initial in-equality: women as secondary, men as primary.

In each of Conrad's 14 novels the main character is male. This shows a disparity already. But a book with one main character is limited to making just one choice about that character's gender. I wanted to find a better metric for counting the balance of male and female characters in a novel.

After deliberating over more complex methods, I settled on a simple one: the ratio of uses of *he* compared to uses of *she*. It's not perfect, but I think it can give you a sense of the gender skew in any book. The count of *he* versus *she* gives a rough breakdown of how the actions, thoughts, and descriptions of male characters match up against female characters.

For example, look at *The Hobbit*. Tolkien used *he* just under 1,900 times in the book. How many times did he use *she*? Once. It was toward the beginning, referring to Mrs. Bilbo Baggins. If you've read *The Hobbit* it's not a stretch to say that it's 99.9% male. Everyone we see—all the elves, dwarves, hobbits, and even the birds—is male.

Joseph Conrad, who wrote his novels at the start of the twenti-eth century, also skewed male. In all 14 of his novels Conrad used the word *he* three times more often than *she*. For every three occa-sions wherein he described the actions, thoughts, or qualities of a man, he described a woman just once.

I decided to cast a wider net, looking at all 100 books on our classic literature list. The following chart shows the *he* versus *she* percentage of each book. If, like Conrad, an author uses *he* three times for every *she*, the corresponding percentage would come out to 75%. Books written by men are represented by black bars. Books written by women are represented by gray bars. Numbers on the right are rounded to the nearest percentage point. If you take a minute to look through, it's clear that the books with the most female mentions are written by women and the books with the most male mentions are written by men.

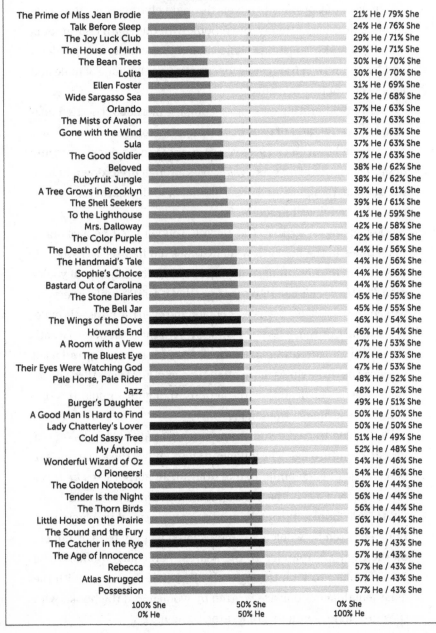

Comparing the Use of *He* vs. *She* in Classic Literature

The Prime of Miss Jean Brodie	21% He / 79% She
Talk Before Sleep	24% He / 76% She
The Joy Luck Club	29% He / 71% She
The House of Mirth	29% He / 71% She
The Bean Trees	30% He / 70% She
Lolita	30% He / 70% She
Ellen Foster	31% He / 69% She
Wide Sargasso Sea	32% He / 68% She
Orlando	37% He / 63% She
The Mists of Avalon	37% He / 63% She
Gone with the Wind	37% He / 63% She
Sula	37% He / 63% She
The Good Soldier	37% He / 63% She
Beloved	38% He / 62% She
Rubyfruit Jungle	38% He / 62% She
A Tree Grows in Brooklyn	39% He / 61% She
The Shell Seekers	39% He / 61% She
To the Lighthouse	41% He / 59% She
Mrs. Dalloway	42% He / 58% She
The Color Purple	42% He / 58% She
The Death of the Heart	44% He / 56% She
The Handmaid's Tale	44% He / 56% She
Sophie's Choice	44% He / 56% She
Bastard Out of Carolina	44% He / 56% She
The Stone Diaries	45% He / 55% She
The Bell Jar	45% He / 55% She
The Wings of the Dove	46% He / 54% She
Howards End	46% He / 54% She
A Room with a View	47% He / 53% She
The Bluest Eye	47% He / 53% She
Their Eyes Were Watching God	47% He / 53% She
Pale Horse, Pale Rider	48% He / 52% She
Jazz	48% He / 52% She
Burger's Daughter	49% He / 51% She
A Good Man Is Hard to Find	50% He / 50% She
Lady Chatterley's Lover	50% He / 50% She
Cold Sassy Tree	51% He / 49% She
My Ántonia	52% He / 48% She
Wonderful Wizard of Oz	54% He / 46% She
O Pioneers!	54% He / 46% She
The Golden Notebook	56% He / 44% She
Tender Is the Night	56% He / 44% She
The Thorn Birds	56% He / 44% She
Little House on the Prairie	56% He / 44% She
The Sound and the Fury	56% He / 44% She
The Catcher in the Rye	57% He / 43% She
The Age of Innocence	57% He / 43% She
Rebecca	57% He / 43% She
Atlas Shrugged	57% He / 43% She
Possession	57% He / 43% She

100% She	50% She	0% She
0% He	50% He	100% He

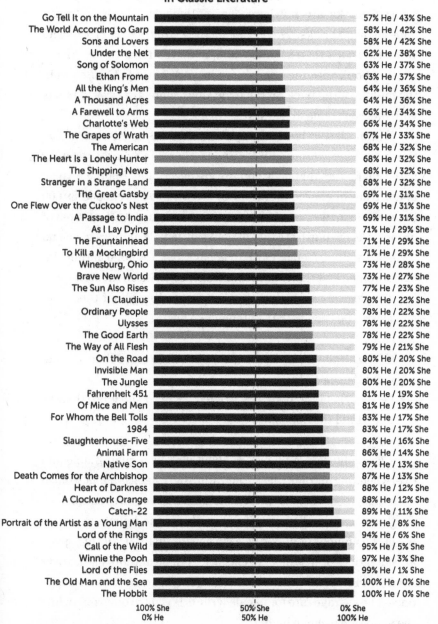

Comparing the Use of *He* vs. *She* in Classic Literature

Go Tell It on the Mountain	57% He / 43% She
The World According to Garp	58% He / 42% She
Sons and Lovers	58% He / 42% She
Under the Net	62% He / 38% She
Song of Solomon	63% He / 37% She
Ethan Frome	63% He / 37% She
All the King's Men	64% He / 36% She
A Thousand Acres	64% He / 36% She
A Farewell to Arms	66% He / 34% She
Charlotte's Web	66% He / 34% She
The Grapes of Wrath	67% He / 33% She
The American	68% He / 32% She
The Heart Is a Lonely Hunter	68% He / 32% She
The Shipping News	68% He / 32% She
Stranger in a Strange Land	68% He / 32% She
The Great Gatsby	69% He / 31% She
One Flew Over the Cuckoo's Nest	69% He / 31% She
A Passage to India	69% He / 31% She
As I Lay Dying	71% He / 29% She
The Fountainhead	71% He / 29% She
To Kill a Mockingbird	71% He / 29% She
Winesburg, Ohio	73% He / 28% She
Brave New World	73% He / 27% She
The Sun Also Rises	77% He / 23% She
I Claudius	78% He / 22% She
Ordinary People	78% He / 22% She
Ulysses	78% He / 22% She
The Good Earth	78% He / 22% She
The Way of All Flesh	79% He / 21% She
On the Road	80% He / 20% She
Invisible Man	80% He / 20% She
The Jungle	80% He / 20% She
Fahrenheit 451	81% He / 19% She
Of Mice and Men	81% He / 19% She
For Whom the Bell Tolls	83% He / 17% She
1984	83% He / 17% She
Slaughterhouse-Five	84% He / 16% She
Animal Farm	86% He / 14% She
Native Son	87% He / 13% She
Death Comes for the Archbishop	87% He / 13% She
Heart of Darkness	88% He / 12% She
A Clockwork Orange	88% He / 12% She
Catch-22	89% He / 11% She
Portrait of the Artist as a Young Man	92% He / 8% She
Lord of the Rings	94% He / 6% She
Call of the Wild	95% He / 5% She
Winnie the Pooh	97% He / 3% She
Lord of the Flies	99% He / 1% She
The Old Man and the Sea	100% He / 0% She
The Hobbit	100% He / 0% She

100% She / 0% He   50% She / 50% He   0% She / 100% He

But saying most authors just prefer to write about their own gender would be an oversimplification. First, the most female-focused books are nowhere near as lopsided as the extreme male-focused books. *The Prime of Miss Brodie* was 21% *he* and 79% *she*. That's the extreme example on the female side. Meanwhile, a book with the opposite split—79% *he* and 21% *she*—is in the middle of the pack on the male side. There are *twenty* books with more extreme male ratios.

Within classic literature by men, *she* was used over 48,000 times, while *he* was used 108,000 times. There is a huge discrepancy in the characters that male authors are describing. But the reverse is not true. In classic literature by women, *she* was used 89,000 times while *he* was used 90,000 times. The near identical rates of pronoun usage illustrate that in books by female authors, men and women are described at close to equal rates. Yet male authors include women less than half as often as they write about men.

The degree to which authors prefer writing about their own gender can be seen with the breakdown below.

- Of the 50 classic books by men, 44 used *he* more than *she* and 6 did the opposite.
- Of the 50 classic books by women, 29 used *she* more than *he* and 21 did the opposite.

Classic literature by men is about men by a quantifiable and overwhelming margin. Classic literature by women is about women more than men, but it's within a short distance of an even split.

The chart on the opposite page shows the imbalance, looking at the number of books by each gender that fall within a given *he*:*she* ratio. For instance, 13 classics by male authors use *he* over *she* between 80–90% of the time while just one classic by a woman does. The average rate for female writers is right around 50% while the male rate is much higher.

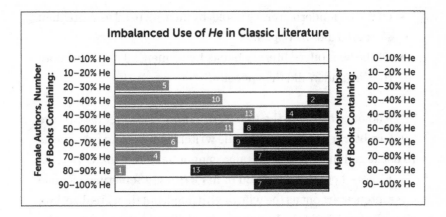

Imbalanced Use of *He* in Classic Literature

At this point you might be thinking that much of the disparity could be a result of the era when these books were written. Sure, Joseph Conrad wrote three times as much about men than women, but he wrote more than 100 years ago.

It's also notable that these "classics" were determined by rankings done by select individuals at organizations like the American Library Association and publications like *Library Journal*. Not every list was transparent in the selection process, so we don't know the gender makeup of the people ranking the books. And even if you were to assume there was no existing bias today, books from the female perspective could have had more difficulty gaining "classic" status in the early twentieth century because of a bias then. If they were not critically popular in their day, the books would have to overcome much more to be on the minds of any literary scholar today.

However, looking at other selections of books that are contemporary, that are not just curated by a small group of individuals retrospectively, the trend is strikingly similar.

- Of the 50 recent *New York Times* bestsellers by men, 45 used *he* more than *she* and 5 did the opposite.
- Of the 50 recent *New York Times* bestsellers by women, 17 used *she* more than *he* and 33 did the opposite.

- Of the 50 modern literary books by men, 42 used *he* more than *she* and 8 did the opposite.
- Of the 50 modern literary books by women, 23 used *she* more than *he* and 27 did the opposite.

And once again, in the *New York Times* bestseller list and the modern literary sample, no female writer ever went so extreme as to use less than 20% *he*. The same cannot be said of male writers.

Elmore Leonard once said of his writing, "Sometimes female characters start out as the wife or girlfriend, but then I realize, 'No, she's the book,' and she becomes a main character. I surrender the book to her." However, digging into the data, I don't think Leonard did as well living up to this idea as he thought. Leonard wrote 45 novels. In not one book did he write *she* more than *he*.

This does not mean Leonard did not write a few strong, original female characters, but in each of 45 cases the book was male dominated. You might be familiar with Leonard's *Rum Punch* (or its Tarantino adaptation *Jackie Brown*), which features the protagonist Jackie Burke. She might be an unforgettable female lead, but since the book has a *he/she* split of two to one she's very much still living in a "man's world," as all of Leonard's novels are.

I don't mean to single out Leonard here. Many other successful writers have written only books that have a male focus. I imagine this list could be as long as you wanted it to be if you kept searching, but by my count, it includes at least the following: Joseph Conrad (14 of 14 novels), Theodore Dreiser (8 of 8), William Faulkner (19 of 19), F. Scott Fitzgerald (4 of 4), Ernest Hemingway (10 of 10), James Joyce (3 of 3), John Steinbeck (19 of 19), Kurt Vonnegut (14 of 14), Salman Rushdie (9 of 9), Jack London (20 of 20), William Gaddis (5 of 5), Elmore Leonard (45 of 45), Jonathan Franzen (4 of 4), Charles Dickens (20 of 20), Michael Chabon (7 of 7), John Cheever (5 of 5), Herman Melville (9 of 9), Cormac McCarthy (10 of 10), and Ray Bradbury (11 of 11).

It's harder to find anyone who does the opposite. Willa Cather, Toni Morrison, Ayn Rand, Edith Wharton, Alice Walker, Gillian Flynn, Virginia Woolf, Charlotte Brontë, Zadie Smith, Agatha Christie, and Jennifer Egan have all written at least one book using *he* more than *she*. Jane Austen, author of six novels, including *Pride and Prejudice*, is the one writer I could find who never wrote a book with *he* more than *she*.

You might be familiar with the Bechdel test, a checklist test to determine if a work of fiction (most often, a movie) shows gender bias. The requirements, on paper, sound simple. In order to "pass" the test, the work must (A) include at least two women, (B) who talk to each other, (C) about something other than a man. The website bechdeltest.com tabulates movies according to the Bechdel test and, as of my writing this, lists 220 movies that were released in 2014. Of these 220 movies a total of 91 failed.

The *he:she* ratio in novels is revealing enough to unmask bias as well, and I think a rule can be built upon it that shows whether a given book skews too far one way or the other. Inspired in part by the Bechdel test, this metric is meant to be a firm answer to the question of whether a novel has a clear gender imbalance. This *he:she* ratio is also better fitted to a novel than the Bechdel test. Books, unlike films, aren't necessarily constructed with a series of scenes with different combinations of characters. Using the pronoun ratio is a singular check that can be calculated in an instant.

My test is simple: If a novel describes male actions three times as much as female actions, it fails the quantitative test. If a book describes female actions three times as much as male actions, it fails also.

The three-to-one (or 75%) barrier is an arbitrary cutoff. The lopsided ratio is in line with Conrad's extreme imbalance. Knowing this ratio, it feels unsettling to pick up a book and be fully aware that for every three actions or descriptions of a man there will be just one mention of a woman (or vice versa).

Many great books fail this test. I'm aware of that. The chart on pages 42–43 shows the number of books which skew past the 25–75 split one way or another. Two classics, both by women, would fail for being too female heavy (*The Prime of Miss Brodie* by Muriel Spark and *Talk Before Sleep* by Elizabeth Berg). Meanwhile 27 classics exceed the 75% male threshold. Twenty-four are by men and just three (*Death Comes for the Archbishop* by Willa Cather, *The Good Earth* by Pearl Buck, and *Ordinary People* by Judith Guest) are penned by women.

I realize many people reading this might believe that a quantitative *he:she* test holds no bearing on the success of a novel. *The Old Man and the Sea* fails. It's a story with just a few characters. Other than the old man, fishermen, and a marlin there isn't much else. With 99% *he*, it fails the test hard.

But it should be noted that *The Old Man and the Sea* would not come close to passing the Bechdel test either. Does that mean the book must be considered sexist? No. It's a book in part about isolation, so the lack of interaction is important. A book can fail either test and still be great, but there should be a justification.

For many of us, novels are a portal, a way of exploring the broader world and understanding how people act within it. We live in a world where one in two people are women. There's no reason to think that every novel must be in lockstep with this ratio, especially if the setting is unique. But if you are a reader and every book you read doesn't even achieve a one-in-*four* ratio, chances are you're not getting a true reflection of, or gaining a true appreciation for, how other people act in the world.

Popular crime writer P. D. James has said, "All fiction is largely autobiographical and much autobiography is, of course, fiction." Word frequencies can't help us explore the second part, but the first half of her quote, that all fiction is autobiographical, deals with the notion that all writers, consciously or unconsciously, write characters based on some part of themselves. In exploring this possibility, word frequencies offer a portal into an author's mind.

Her theory, however playfully it was meant, does help explain why the gender balance of male authors is so skewed toward male characters. Writing characters based on yourself does have one advantage: It gives writers a chance to write about what they know.

However, I can't imagine any writer advising a novelist to only include characters that are based on their own personality. If anything, the skill of a great writer is to create good, believable characters from different backgrounds and with different motivating forces. Gender is one of the biggest dividing lines between characters and one of the trickier challenges for some writers. Writers need to make characters of the opposite sex believable, which means needing to engage with cultural norms and vocabularies that meet with readers' general expectations. At the same time, playing into stereotypes or oversimplifying is the quickest way to turn off readers from continuing.

Looking at how writers describe characters of the opposite sex provides us with a new way of understanding the choices, and unwitting assumptions, that factor into how we see other people—and who we see on the page.

Let's start by looking at the word *scream*.

In the top 100 classic literature books, a form of *scream* appears after the word *he* or *she* a total of 158 times. For example, at the end of *The Grapes of Wrath*, when Rose is escaping from an impending flood, Steinbeck writes, "And Rose of Sharon had lost her restraint. **She screamed** fiercely under the fierce pains." If we look at all the instances where male writers used the word *scream*, it is used twice as often after *she* than *he*. In other words, male writers make their female characters scream more often than their male characters.

But that's not enough to say that male authors have gone rogue. For if you look at the use of *scream* by female authors, the result holds at an almost identical rate. In other words, female writers also make their female characters scream more often than their male characters.

The graphic below shows the usage rate of *screamed* (or *screams*) when it follows either *he* or *she*. "She screamed" was used at a rate of 6.0 for every 10,000 appearances of the word *she* in texts by male authors and 7.0 per 10,000 by female authors. Meanwhile "he screamed" was used at a rate of 3.8 for every 10,000 appearances of *he* in texts by female authors and 2.9 per 10,000 by male authors.

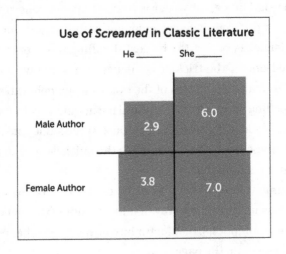

The number of instances of *scream* is not huge, but it's still large enough to draw a conclusion: Male and female authors both are more apt to describe their female characters as screaming than their male characters. I also looked at the sample groups of recent *New York Times* bestsellers and modern literary fiction to see if there was any change over time or genre, and the pattern stayed the same. In these groups, *screamed* is used 50–100% more often to describe female characters than male, and this holds for authors of both genders.

Just as there are words that tend to be paired with female characters, there are those that latch on to male characters. And while men might not be screaming as often, they sure do like to grin. Below is a chart in the same style for all the appearances of *grinned* following *he* or *she* in classic literature.

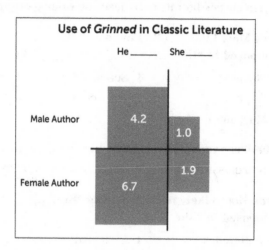

And in modern popular and literary fiction, authors of different eras and styles are consistent—though there's a lot more grinning going on in popular fiction.

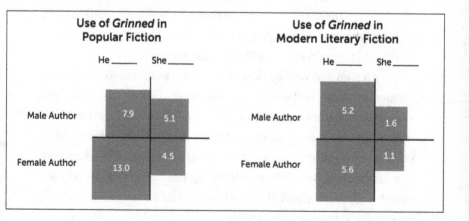

The patterns in the charts show that these word choices hold across time or genre. It was not my instinct going in that *grinned*

was a gendered verb in any form—but the data reveals an undeniable trend. Men *grin*.

Below are the top five words, like *screamed*, that are used most often in classic literature to describe women over men.

Words Most Likely to Be Found as "She _____"
as Opposed to "He _____"

1. Shivered
2. Wept
3. Murmured
4. Screamed
5. Married

And here are the top five words, like *grinned*, that are used most often in classic literature to describe men over women.

Words Most Likely to Be Found as "He _____"
as Opposed to "She _____"

1. Muttered
2. Grinned
3. Shouted
4. Chuckled
5. Killed

Women *murmur* yet men *mutter*. Men *shout*; women *scream*. Women neither *grin* nor *chuckle*, but *smile* is more likely to follow *she*. Each of these trends holds across recent popular and literary fiction, with one interesting reversal: In modern literary fiction, *married* is more common after *he* than *she*.

Both men and women describe men as killing more often than women. And this is the only word in any of these lists that we can check against real-world data. Statistics show that men commit 90% of all murders. Government agencies, however, do not keep stats on how many *grin*s or *chuckle*s are committed by each gender. Regardless, it seems like these words have come to be connoted with gender and that both male and female authors have picked up on those connotations.

But what if we took this a step further? What if there are words that men use to describe women, but which a woman would never use to describe herself or another woman? These are the words that could highlight the biggest differences in how we view the world. If you are an author writing about a character of the opposite gender, what makes that character believable or real? You want to make sure that you're describing their thoughts and actions in a way that reflects how *they* see the world, using the language *they* would use. Otherwise, the illusion can shatter.

One word that fits this description is *interrupted*. It's not the most common word in any writer's works, but especially in classic literature it is used much more commonly in reference to female characters when the author is male.

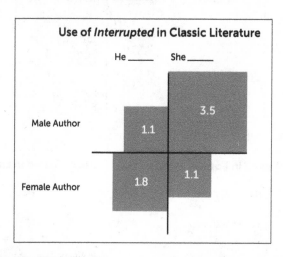

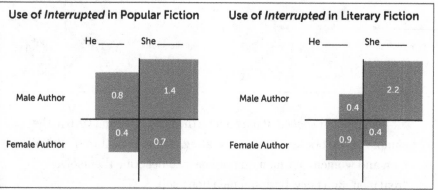

There are also a handful of words, in the sample of books examined, that authors rarely invoke when describing the opposite sex. While men described all their characters as having fear, women assigned *fear* to their male characters significantly less often. See the chart below:

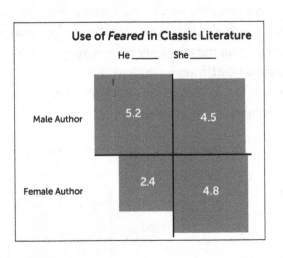

Use of *Feared* in Classic Literature

He _____ She _____

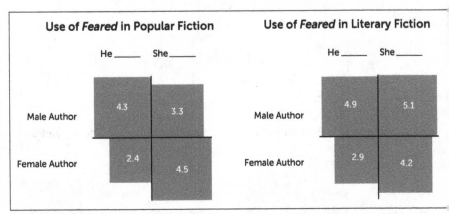

Use of *Feared* in Popular Fiction

Use of *Feared* in Literary Fiction

Or consider *sobbed*. It may not be the most common verb in the sample of 300 books, but it is revealing. Women use it to describe men and women, but men do not use it to describe themselves. If "real men" don't cry, fictional men don't sob.

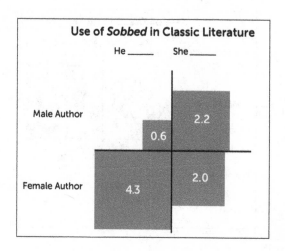

Use of *Sobbed* in Classic Literature

He _____ She _____

	He	She
Male Author	0.6	2.2
Female Author	4.3	2.0

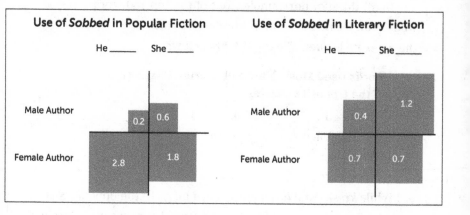

Use of *Sobbed* in Popular Fiction

He _____ She _____

	He	She
Male Author	0.2	0.6
Female Author	2.8	1.8

Use of *Sobbed* in Literary Fiction

He _____ She _____

	He	She
Male Author	0.4	1.2
Female Author	0.7	0.7

Then, perhaps most interesting of all, there are the words that both sexes give to the opposite gender. Male authors describe their female characters as *kissing* at a higher rate than their male characters. Female authors do the opposite, describing their male characters kissing more often.

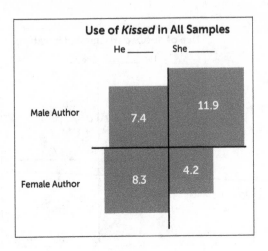

In all three of our samples, out of the top 150 words, *kissed* was the single most common word used to describe characters of the opposite gender. The top five are below:

Words Used Most Often to Describe Characters of the Opposite Gender

1. Kissed
2. Exclaimed
3. Answered
4. Loved
5. Smiled

While *kissed* and *loved* and *smiled* all go to the opposite sex, consider the use of *hated*. The h-word is used most often in classic literature to describe characters of the author's own gender.

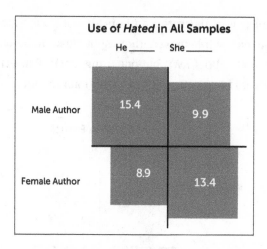

Use of *Hated* in All Samples

He _____ She _____

	He	She
Male Author	15.4	9.9
Female Author	8.9	13.4

Here are the top five words used to describe characters of the author's own gender:

Words Used Most Often to Describe Characters of the Same Gender

1. Heard
2. Wondered
3. Lay
4. Hated
5. Ran

Trying to draw too much meaning out of these findings is a bit like reading tea leaves. But I don't think it's unreasonable to speculate that some of the words writers used to describe the opposite sex, such as *loved* and *kissed*, serve as a kind of wish fulfillment. It may be a leap to base a theory of love on 300 novels, so I decided to test it on a bigger data pool as well. I downloaded more than 40,000 Literotica.com stories in the "Erotic Couplings" section and found a similar pattern with the word *kiss*.

I combined *I kissed* and *He kissed* for all male authors and *I kissed* and *She kissed* for female authors—to avoid missing first-person stories. And following is the rate for each combination. There's a huge asymmetry, but not one that's identical to the other samples. Female erotica authors almost always attribute the kissing to a man, while male authors split closer to 50-50. I'll leave you to speculate about *why*, but one thing that's clear is that in the realm of sexual fantasy people's imaginations are not fully aligned.

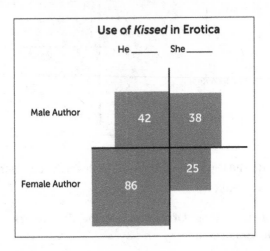

All the examples in this section show how male and female writers describe the world, and their male and female characters, in different ways. It's intriguing from a psychological angle, but these findings are also worth keeping in mind for authors who are trying to capture characters of all genders and trying to appeal to a wide range of readers.

After all, you don't want all your characters to be clones of common stereotypes or of your own persona. Writing what you know is important, but not considering the perspective of others can lead to the downfall of a work. As science fiction author Joe Haldeman has quipped, if there's one thing the "write what you know" default has resulted in, it's a glut of "mediocre novels about English professors contemplating adultery." Let's keep expanding the literary imagination.

Chapter 3

Searching for Fingerprints

Your style is an emanation from your own being.
—KATHERINE ANNE PORTER

The whodunit is not limited to the world of crime; it's also a staple of literary scholarship. A book lands with a thud on an editor's doorstep one morning, with no clues to its origins. It's anonymous, pseudonymous, unattributable—yet unignorable.

Who wrote it? Interested critics might have their favorite suspects. Opportunistic writers may even quarrel over credit. But the answer, as with any mystery, lies in the cold, hard facts. Which is to say, aspiring literary detectives will need to turn to the numbers.

Let's return to *The Federalist Papers* controversy from the introduction, one of the most famous literary mysteries solved in the past century. In order to urge ratification of the Constitution in the late 1780s, James Madison, Alexander Hamilton, and John Jay each wrote essays that appeared in New York newspapers under the pseudonym "Publius." Between them, the three men wrote a total of 85 essays, but no one took credit for any individual essays until decades later. When Madison and Hamilton outlined who wrote each essay, there was a contradiction.

Twelve of the essays were claimed by both Madison and Hamilton.

In 1963 two statistics professors, David Wallace and Frederick Mosteller, put forward evidence in *Inference in an Authorship Problem* that would end the near two-century-long debate. Their probabilistic case was objective and detailed. It quantified writing styles. It succeeded where qualitative arguments had suffered.

Their biggest step forward was treating words like random variables. Instead of viewing the words as sacred they looked at them the same way they would study the rolls of a die or a flip of a coin. The two looked at the frequency of hundreds of words, which was not easy to do in 1963. They took copies of each essay and dissected them, cutting the words apart and arranging them (by hand) in alphabetical order. At one point Mosteller and Wallace wrote, "during this operation a deep breath created a storm of confetti and a permanent enemy."

In particular, they started looking at a handful of words that were used by one author but not the other. In his known papers Alexander Hamilton used the word *while* but never the word *whilst*. Madison used the word *whilst* but not *while*. The professors listed the rate of *enough, while, whilst,* and *upon* per 1,000 words in the Hamilton, Madison, and disputed papers.

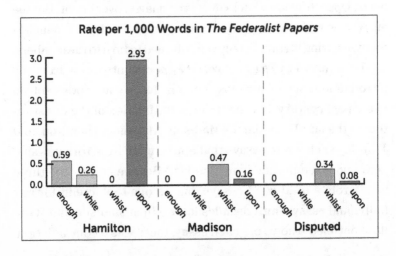

The graph of Mosteller and Wallace's figures on the previous page lends itself to an easy conclusion. Hamilton used *enough* and *while* but Madison and the disputed papers never did. Hamilton used *upon* to a high degree, but Madison and the disputed papers used it at a much lower rate. *Whilst* is absent from Hamilton's writing but present in the disputed papers. It looks like it can't be Hamilton, right?

But this was not enough for Mosteller and Wallace. It was just four words. If that's all you saw you might think there is no reason for more data or more analysis. However, if Mosteller and Wallace had looked at *according*, *whatever*, *when*, and *during* they would have found the opposite:

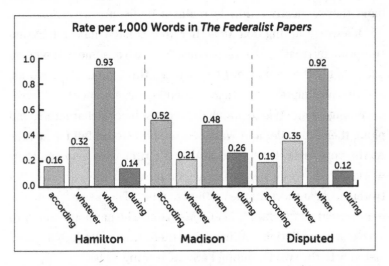

The graph above makes the disputed papers line up with Hamilton's patterns. I had to search through hundreds of words to find numbers this contradictory, but the point remains that not every word's frequencies are constant in every text. Most words don't line up perfectly for either Madison or Hamilton—the eight you see here are rare. And if you look through enough words, you'll be able to find a handful that can support any conclusion: Hamilton, Madison, even you or me.

That's why Mosteller and Wallace created a system to weigh the importance of a large number of factors. The exact details rely on some equations that we don't need to get into here. But the thought process is straightforward. Each word allowed them to make a small calculation about who the likely author was. When the differences in word frequencies were all combined, the outliers cancelled out. All of those small probability calculations, when multiplied together, amassed to a rock-solid prediction: A text with this level of *the* usage, that level of *during* usage, that level of *whatever* usage, etc., would never in a thousand years have been written by Hamilton. It would have taken an outright miracle, a sudden change to every marker of his writing style, for Hamilton to pen those 12 essays. On the other hand, they sat neatly within the realm of Madison's own style.

It's worth noting that Mosteller and Wallace make a huge assumption by treating words like dice. The two assume that writers use roughly the same word frequencies throughout their works, and this assumption is critical to their equations' success. If writers change their style to match different subjects, characters, and plots, then Mosteller and Wallace's method would fail frequently. At the very least, the variation that a writer uses between their works needs to be insignificant compared to the variation between other authors for the method to work. That assumption ultimately held up for Hamilton and Madison: The method's ability to arrive at a prediction confirms that there was an underlying consistency to the two Founding Fathers' writing styles.

But I've long wanted to see just how far the theory can go— to test whether something like a literary fingerprint exists for famous writers.

The rest of this chapter will look in depth at Mosteller and Wallace's assumption that word choice is constant. If it is correct, and style does not change from book to book, then their method should work just shades off 100% of the time, regardless of genre. Forensic scientists are able to use fingerprints to identify people because the ridges on people's fingers do not change. But are the stylistic fingerprints that each author leaves in their writing unique enough, and permanent enough, for Mosteller and Wallace to pick them up without fail?

Testing Mosteller and Wallace on Fiction

The uniqueness of fingerprints has been known for thousands of years. No two fingerprints are the same, and civilizations as early as ancient Babylon and China used them to ensure contracts.

Fingerprints don't tell you anything about the suspect by themselves. The identification process only works if you have a set of the suspect's fingerprints on file or a database to compare against an unknown print. What if the same could be done for books? Mosteller and Wallace's method suggests that writers have a hidden fingerprint, too: Authors leave a pattern of words wherever they write. And in the last two chapters we've assembled quite a few samples.

To start experimenting with this idea, I gathered a mixed collection of great and popular books, almost 600 books by 50 different authors. This would serve as my full database. (The full list is included in the Notes section on page 264.) Then I chose one book, *Animal Farm* by George Orwell, and removed it from the sample.

Mosteller and Wallace didn't build their method specifically for novels. And though people will sometimes attempt to identify one particular book, no one has ever gone through and replicated the professors' original methods on a large set of novels by known authors. To find out if it could work, I started with a small test.

First I set *Animal Farm* as the unknown fingerprint. I then treated Hemingway's ten novels and Orwell's five other books as my known sample. With two possible options, Mosteller and Wallace pinpointed Orwell as the author of *Animal Farm*. It was a good start, but a coin would have a 50-50 chance of being accurate after one test.

Then, I expanded the list of candidates. One by one I set my computer to test *Animal Farm* against each of the other 48 authors in the sample. This includes authors considered among the greats, such as Faulkner and Wharton. It features many writers who have found huge popularity, such as Stephen King and J. K. Rowling. And it includes a handful of other writers who have achieved recent literary success, such as Jonathan Franzen and Zadie Smith. For each author I included their complete bibliography of novels. In each of the 48 test cases the result was the same: Mosteller and Wallace were able to correctly identify Orwell as the author of *Animal Farm*.

I wanted to see if this was a fluke. Perhaps *Animal Farm* was an outlier with a weird style that had unusual results. I compared each of Orwell's other five books (*Burmese Days, A Clergyman's Daughter, Keep the Aspidistra Flying, Coming Up for Air*, and *1984*) to the other 49 writers in the sample. Each time, I removed the book in question from Orwell's sample and treated it as an unknown text. Out of 245 comparisons using Mosteller's system, it was right 245 times. In every case it listed Orwell as the more probable author.

I then expanded further, testing every single book in the sample, pitting each one head-to-head against its actual author and each of the 49 other authors. This totaled 28,861 tests. I figured it would be the best way to confirm if Mosteller and Wallace has validity on long fiction.

Every time, the method was looking at the same basic 250 words. Of the almost 29,000 tests Mosteller's system worked all but 176 times. This is over a 99.4% success rate.

How is it possible that a system so simple works so well?

The reason it works is that authors *do* end up writing in a way that is both unique and consistent, just like an actual fingerprint is distinct and unchanging.

Consider Khaled Hosseini, Zadie Smith, and Neil Gaiman. They do not write about the same subjects or with the same tone, but they are all modern-day popular authors with overlapping international audiences. Mosteller and Wallace can distinguish their work with 100% accuracy (28 out of 28) by looking only at 250 simple words. In fact, even just looking at *the* and *and*, the two most common words in the sample, you can see distinctions among the three writers. Take a look at the graph below.

The "fingerprint" of *the* and *and* is illuminating. If one datapoint's label were removed from this chart, we'd have little trouble predicting the author based on where it falls. With the simple eyeball test you could guess right the majority of the time using only the *two most common words*.

There are exceptions of course. The most obvious is *Anansi Boys*. It is the one book (asterisked below) by Gaiman with fewer than 500 *the*s per 10,000 words. This looks like it could be categorized as Hosseini or Smith before Gaiman.

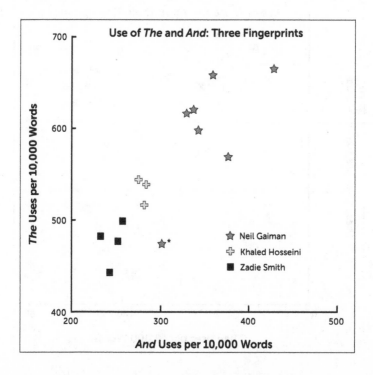

But Mosteller and Wallace has something going for it. *The* and *and* are just a fraction of the words used to distinguish texts. On the sample of 50 writers, using Mosteller and Wallace to predict authorship with just the word *the* is correct in 71% of head-to-head comparisons. With *the* and *and* it's right 83% and with the top ten most common words it gets by at 96%.

Though writers may have a book with indistinguishable or out-of-character patterns for a single word, by the time the couple hundred most common words are accounted for, the style is undeniable. Consider *these* and *then*, which when graphed reveal a distinct Gaiman cluster. *Anansi Boys*, which was out of character on the *the* and *and* plot, is asterisked again. This time, it's right in the middle of Gaiman's other works.

The method is not entirely perfect. Of every comparison, Wil-

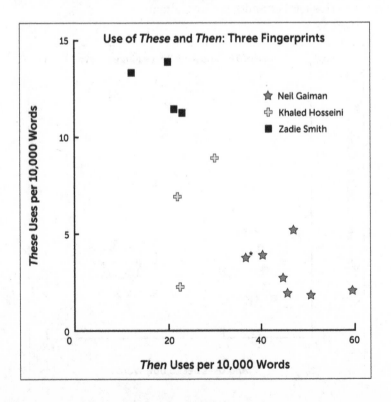

Use of *These* and *Then*: Three Fingerprints

liam Gaddis's *The Recognitions* was the most misidentified novel, with 39 out of 49 authors coming up as the more likely author than Gaddis. Three out of nineteen Steinbeck novels listed Mark Twain as the more probable author. But with a failure rate of just one for every 165 head-to-head tests, Mosteller and Wallace's system works wonders.

The Magic of Probability

The previous section showed that Mosteller and Wallace worked 99.4% of the time on known works, but what happens when a writer is actively trying to disguise themselves? The central assumption of the model is that writing style is constant, but can an author stay incognito by trying to write for a different fan base or in a different genre?

Consider the cases of Richard Bachman and Robert Galbraith.

Richard Bachman is a horror writer. For years he ran a dairy farm in New Hampshire and wrote at night. His life was tragic. Bachman's only son drowned in a well and the author himself died of cancer in 1985. Fortunately for his readers, he left behind a large volume of works that are still being published to this day.

Richard Bachman is also alive and well. He is a pen name of Stephen King.

The true identity of Bachman was unmasked when a reader noticed similarities between the style of Bachman's writing and another of his favorite suspense writers. He did a search of the Library of Congress catalog and found the book listed under, just as he'd suspected, Stephen King. The master of mystery novels had failed to cover his tracks.

But could Mosteller's formula have detected Bachman's true identity from the text of his novels alone?

The simple answer is no. It can be used to detect if the true author is writer A or writer B when A and B are both known. In the case of Bachman the alternative was that Bachman was real, or at the least a separate unpublished author. There would have been no way to tell with any certainty that King was the author.

However, what if that industrious reader in 1985 had decided to take the investigation into his own hands and replicate Mosteller on Bachman with a sample of bestselling authors? Who was more probable to be Bachman? Agatha Christie or James Patterson? Elmore Leonard or Tom Wolfe? Or Stephen King?

These tests could show distinct similarities or differences, even if they couldn't catch the true author red-handed. If King and Bachman turned out to have little in common by the numbers, then Mosteller and Wallace could at least dissuade you of your pet theory.

For all four of Bachman's books, when compared to our fifty top authors, Stephen King comes up as number one every time. That's 196 correct identifications out of 196. Of course, many of these pairings seem trivial. Charles Dickens would not be confused for a horror novelist by anyone. But the success is still lopsided enough that it could have added firm confidence to the reader who noticed the qualitative similarities.

Following are the ten authors who were top five most probable and least probable.

Most Probable to be Richard Bachman

1. Stephen King
2. James Patterson
3. Tom Wolfe
4. Gillian Flynn
5. Neil Gaiman

Least Probable to be Richard Bachman

1. Suzanne Collins
2. J. R. R. Tolkien
3. Veronica Roth

4. E L James
5. Jane Austen

Not all pseudonym speculations turn out to be true. In 1976 American radio host John Calvin Batchelor forwarded one of the more far-out literary conspiracy theories I've heard. In *SoHo Weekly* he wrote:

> What I am arguing . . . is that J. D. Salinger, famous though he was, simply could not go on with either the Glass family, which had by 1959 his weight to bear, or with his own nationally renowned reputation . . . So then, out of paranoia or out of pique, J. D. Salinger dropped 'by J. D. Salinger' and picked up 'by Thomas Pynchon.'

Since then Batchelor has backed down from his theory. He received a letter from Thomas Pynchon after the article was written saying he was mistaken. The rumor has persisted, even if in jest, as a function of how reclusive both Pynchon and Salinger are or were.

We've seen Mosteller's math work well on *The Federalist Papers* and Stephen King. What does it say about Pynchon and Salinger?

Again, we would not be able to definitively confirm the theory that Salinger and Pynchon are the same person, but the empirical evidence here can rule out that Salinger and Pynchon are the same person.

I compared Salinger's work (excluding short stories, so just *The Catcher in the Rye* and *Franny and Zooey*) against 49 other authors. Combined with Pynchon's eight books, this amounted to 392 different tests. In 42 of these tests it identified Salinger as the more probable author. For instance, J. D. Salinger was more probable to be the writer of Pynchon's *Inherent Vice* than Ernest Hemingway. But in 350 out of 392 cases, Salinger turned out less likely to be the author.

Quantitatively, then, Salinger's writing bears no similarity to Pynchon's novels on the word-for-word level. The test confirmed what we already know: Pynchon is not Salinger, and radio hosts who put forward attention-seeking theories are more often wrong than right.

There is one more pseudonym challenge that I've wanted to test—one where the author is switching genres. And the perfect example arose when Robert Galbraith arrived on the scene. Like Richard Bachman, Galbraith doesn't actually exist. He's J. K. Rowling's pen name. But whereas King wasn't trying to change his writing much as Bachman, Rowling *was* trying to change her style in the Galbraith books. The Galbraith books are detective novels written for Muggle adults, while the entirety of our Rowling sample consists of the Harry Potter books, full of magic and geared toward young adults. This is a major shift. What if Mosteller had been born fifty years later and decided to investigate Robert Galbraith and J. K. Rowling instead of obsessing over *The Federalist Papers*? Would the change in genre mean a departure in style?

Remarkably, even with the leap out of the Harry Potter universe, Mosteller and Wallace could pick out J. K. Rowling as the best match for all three Galbraith books.

Most Probable to be Robert Galbraith

1. J. K. Rowling
2. Jonathan Franzen
3. Stephen King
4. James Patterson
5. Jennifer Egan

Rowling wrote one detective novel, *The Casual Vacancy*, under her own name, but that wasn't included in my earlier sample. Her Harry Potter books alone were the best match for all three of her Cormoran Strike novels. It was accurate in 147 out of 147 head-to-head tests.

Here's Harry Potter compared to Cormoran Strike as well as the two other most popular detective series (according to a Good reads.com vote), Inspector Gamache by Louise Penny, and Harry Bosch by Michael Connelly. The two words being compared are *but* and *what*.

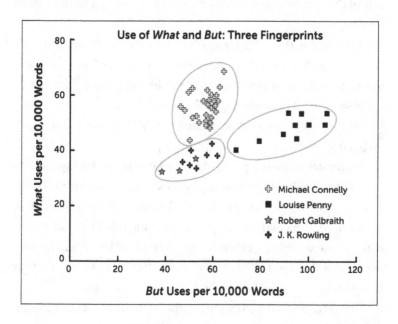

Perhaps there are slight differences among word frequencies from Potter to Cormoran, but when Rowling shifts in writing detective fiction her prose doesn't change at its core. The word frequencies depend more on the writer than the genre. Her writing style stayed closer to the Harry Potter universe than the worlds of Louise Penny or Michael Connelly, and when hundreds of words are taken into consideration (instead of just two) it becomes exceedingly hard for her work to be mistaken for that of many other writers.

Rowling's transformation to detective writer is just one test case, but it's a powerful one. Writers can change genre, and attempt to hide their identity, but that doesn't mean they can hide their writing.

Along Came a Co-author

James Patterson is a prolific writer and his readers are prolific in their consumption of his work. A *New York Times* article on the writer stated that between 2006 and 2010 Patterson was the author of one out of every 17 hardcover novels bought in the United States.

Even since then, Patterson has ramped up production. He started as a thriller writer, publishing around a book a year and now runs multiple series. In 2014 he published 16 books. Patterson has also started to branch off from his thriller roots into fiction geared toward young middle schoolers with his series titled Middle School.

Patterson is quoted as saying, "I believe we should spend less time worrying about the quantity of books children read and more time introducing them to quality books that will turn them on to the joy of reading and turn them into lifelong readers." But it's not as if he has anything against quantity. In all of the 1990s he published a total of ten books, fewer books than he puts out per year these days.

Here is a graph showing the number of books by James Patterson published each year between 1976 and 2014.

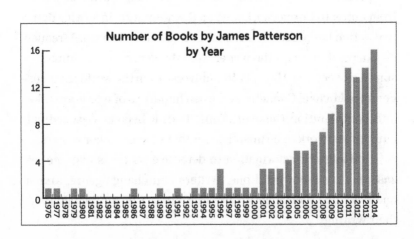

Number of Books by James Patterson by Year

Despite what the pattern of the graph suggests, James Patterson is not on pace to keep writing books at an increasing rate ad infinitum. For one thing, he'd run out of co-authors first.

How does Patterson manage to publish so many books a year? He is not shy about his process. In a *Vanity Fair* profile of Patterson by Todd Purdum, the author said that the way he works with collaborators is to detail an outline. Then the co-authors are responsible for turning the outline into a draft. Here's an excerpt from Purdum's piece of one of Patterson's outline descriptions: "Nora and Gordon continue their quick banter, funny and loving. We like them. They're good together—and not just when they're standing up. A minute later the two engage in some terrific, earth-moving sex. It makes us feel great, horny, and envious." That's a lot of weight left on the co-author's shoulders.

For comparison, below is the number of books by James Patterson *without* a listed co-author.

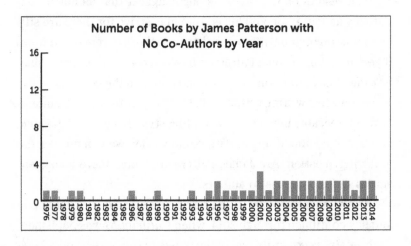

Patterson has four writers with whom he's published at least five novels: Andrew Gross, Howard Roughan, Maxine Paetro, and Michael Ledwidge. These four have worked with Patterson (but not with each other) on a combined 37 novels.

Most of Patterson's co-authors have not published enough independent works to judge against the books they co-authored. However, we can compare these partnerships against one another. If we run the Mosteller test on all of these 37 novels the test is 111 for 111. It recognizes all the books co-written with Andrew Gross, for instance, and can distinguish them from those co-written with Maxine Paetro.

And on the other side of the coin it has a low error rate distinguishing between a Patterson solo project and a Patterson co-write. The word frequency equations were correct 94% of the time (117 times out of 125). It misidentified, for instance, that *Confessions of a Murder Suspect* was a solo project when it was actually co-written with Maxine Paetro. It also misidentified a few books (like *Cross My Heart*) as more similar to the co-written books with Michael Ledwidge even though they are solo books. But on the whole Mosteller and Wallace can tell.

The results on the previous page suggest that as much consistency as Patterson and his editors may strive for there are still major distinguishing differences between the different co-authors. If you are a fan of some Patterson books more than others, it may be time to pay attention to the second name on the cover as well.

Even when writing within a single series, Patterson's co-authors have a noticeable impact on the writing style. Because of the huge number of combinations in Patterson's works, we can answer the following question: Are James Patterson's works more consistent across series or across co-authors?

The Women's Murder Club book series started with *1st to Die* and has continued through 2014, when *Unlucky 13* was published. Andrew Gross co-wrote two of the books in this series while Maxine Paetro co-wrote ten. Both these authors have written other books with Patterson not in the series.

Does Mosteller say Gross's book *2nd Chance* is more similar to other books in the same series co-written with Paetro or more similar to other books co-written with Gross, even if they're in a different series?

The math places *2nd Chance* closer to Gross's *other* works than to Paetro's books in The Women's Murder Club series. If we look at the ten books co-written by Paetro the same is true. Mosteller picks out the co-author even across series.

Without a point of comparison, it's impossible to tell if a Patterson-Gross book is more similar in style to Patterson or Gross. None of the many Patterson co-authors have a sizable library of their own. So although the numbers show there is a clear difference between each co-writer and the co-written books from the solo projects, it's possible that each co-author was just adding a dash of flavor that made them unique.

The burning question that many readers have, however, is whether their favorite writer is using a co-writer or essentially employing a ghostwriter. This line between ghostwriter and co-writer is not always clear or agreed upon. Some people may argue that just because one writer does the outlining and the other writer does the actual writing, that doesn't mean it was ghostwritten. No matter your viewpoint on the distinction, the books—Patterson's and other big-name authors'—are marketed in a way that obscures the roles. Consider the cover here of a book listed as "Tom Clancy with Mark Greaney."

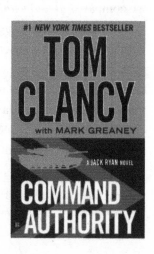

The average reader seeing this mass-market cover in a grocery store would assume that Clancy was the lead writer of the story in every way. Clancy is a huge name, known for his hits like *The Hunt for Red October* and *Patriot Games*. In his career he wrote 13 novels as the sole author. He also co-wrote a number of novels as well as getting involved in "creating" novels. The series

Tom Clancy's Op-Center bears Tom Clancy's name, and he is credited as the "creator." But he wrote none of them; Jeff Rovin did.* For every one book that Tom Clancy authored himself he "created" five others.

When Clancy did co-write, the author he shared a byline with the most was Mark Greaney. They wrote three books together. Greaney has also published five books independent of Clancy. All his collaborations with Clancy are listed as "Tom Clancy with Mark Greaney," even if you have to squint to find Greaney's name on the cover.

If we run Mosteller and Wallace on each author's solo novels, the results are what we would expect. It correctly identifies Clancy's books 13 times out of 13 and Greaney's five out of five. The authors' styles are distinct.

The three books that Clancy and Greaney co-authored were *Command Authority*, *Threat Vector*, and *Locked On*, all novels in the Jack Ryan series. When we run the numbers on these books, however, all three come out Greaney over Clancy. If the disputed documents in Mosteller and Wallace's paper had been the three co-written books instead of the 12 *Federalist* essays, they would pick Greaney every time. Look, for instance, at what we see when we compare *but* and *what*.

The nondisclosure agreements that co-authors sign to work with mega-authors restrict them from revealing how the writing was split up. Without the breakdown of the method, it's hard to get too detailed in the analysis. But to get a more granular look, I split all of the Clancy, Greaney, and "Clancy with Greaney" books into 5,000-word chunks. I then used Mosteller and Wallace methods on each small section. The attribution of the divided books is shown on page 78.

---

* Since Clancy's death in 2013 the series was relaunched after a nine-year break. The new books are written by new authors, Dick Couch and George Galdorisi.

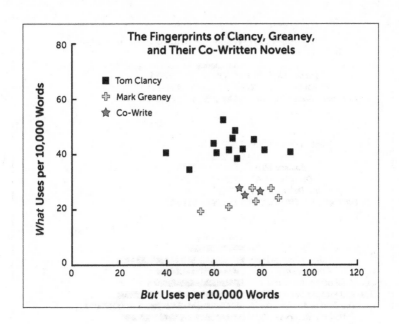

The Fingerprints of Clancy, Greaney, and Their Co-Written Novels

For these short 5,000-word snippets, Mosteller and Wallace is nowhere near the 99% perfection that it achieves on entire novels. We know that because sections in *The Hunt for Red October* are attributed to Greaney despite the fact that he was 16 years old when Clancy published it. Maybe the sections that show up as more Clancyesque in the collaborative books were written by Clancy. Or maybe Clancy wrote around 2,000 of every 5,000-word section, and there are just a few samples that happened by luck to resemble his writing more. In either case, the patterns in the "Clancy with Greaney" books suggest that the co-authorships relied more on Greaney's writing than Clancy's.

In an interview Greaney said that when collaborating with Clancy he "never tried to copy [Clancy's] style," and Mosteller and Wallace bear this out. Greaney's writing style came through much more in the final drafts than Clancy's own. If you loved the plot twists and structure, then you could likely thank both Clancy and

Mosteller-Wallace Attribution of Authorship

Tom Clancy (with Mark Greaney)
- *Locked On* (2011)
- *Threat Vector* (2012)
- *Command Authority* (2013)

Mark Greaney
- *The Gray Man* (2009)
- *On Target* (2010)
- *Ballistic* (2011)
- *Dead Eye* (2013)
- *Support and Defend* (2014)
- *Full Force and Effect* (2014)

Tom Clancy
- *The Hunt for Red October* (1984)
- *Red Storm Rising* (1986)
- *Patriot Games* (1987)
- *The Cardinal of the Kremlin* (1988)
- *Clear and Present Danger* (1989)
- *The Sum of All Fears* (1991)
- *Without Remorse* (1993)
- *Debt of Honor* (1994)
- *Executive Orders* (1996)
- *Rainbow Six* (1998)
- *The Bear and the Dragon* (2000)
- *Red Rabbit* (2000)
- *The Teeth of the Tiger* (2003)

Greaney. But, if you happened to think it was filled with great descriptions and fast-paced sentences, you may be best advised to pick up another Greaney book next.

Team Mosteller or Team Wallace?

To test the breaking point of Mosteller and Wallace I thought long and hard over what the worst literary nightmare for the mathematical model might be. Was there any type of writing that could trip up the equations? After deliberating I came up with the perfect challenge (which perhaps should have been obvious all along): Twilight fan fiction.

In the sections above I looked into the question of genre and writing style, but fan fiction has an element of specificity. The works are not just the same genre or sub-genre, but the same sub-sub-sub-genre. The actual characters stay the same between different authors. All the texts are written within a short window of time. And even more so, the writers are all heavily influenced by the same canonical author.

If Mosteller and Wallace could identify different authors, even when genre has been neutralized, then it seems like it's a good bet to take on any long-form fiction. This, I imagined, was the method's final showdown.

On the website FanFiction.net, the most popular of many fan-fiction websites, people have written more than 1 billion words of Twilight fiction. I chose Twilight for its enormous sample size. Below is a plot of all stories with 60,000-plus words (long enough to be a full novel) dating from Twilight's first release until the end of 2014. In total, there have been 5,000 novel-length Twilight stories posted on FanFiction.net, which I would be comparing to the four novels in the original series. You can see the mounting popularity of fan fiction as the books (marked B1 to B4) came out, and the huge leap immediately after the first movie (marked M1) was released.

Stephenie Meyer wrote 600,000 words in the Twilight series,

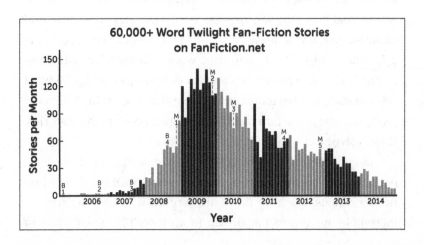

and 153 writers on FanFiction.net have bested her word count in their own Twilight fan fiction.

I ran the Mosteller and Wallace test on Meyer's Twilight books and the top fifty most prolific authors. All these authors, except for Meyer, have written more than 1 million words.

Harkening back to my initial test on *Animal Farm*, I removed one Twilight book at a time and compared that to (A) the other three books in the series and (B) the complete bibliography of each of the fifty fan-fiction writers. No author passed for Meyer. That's a record of 200 for 200.

If you compare all fan-fiction writers, like airedalegirl1, against one another, the results are nearly as strong. Out of all 24,445 combinations of comparing one fan-fiction work to the other fan-fiction authors (or Meyer), the math was right 24,365 times.

The 99.7% success rate is near identical to what we found when looking at writers who varied greatly in genre, era, and subject. If you think that genre is a major tipoff, then Twilight fan fiction would be a major obstacle. Still, Mosteller and Wallace recognize the differences between each author.

I reached out to the top-writing Twilight fan-fiction author of all time, airedalegirl1. I wanted to know how her writing process works (and how long she spends on it). Airedalegirl1, whose real name is Jules, has written 38 stories of 60,000-plus words, totaling 3.7 million words. She is a married woman in her fifties who lives in England. She writes "each day for two to three hours." When I told her she'd written more than anyone else, Jules said, "I've never really thought about how much I've written. I don't plan my stories, they evolve . . . it's just organic."

In addition to sample size, I think airedalegirl1's attitude explains part of the success of Mosteller and Wallace on the fanfiction corpus. Because these are writers who have written an incredible amount of fan fiction in an incredibly small span of time, they are more or less putting words on the page as they think of them. Once they finish one story they start the next. Almost all of these amateur fan-fiction authors have written more words in a few years than professional literary novelists do in a lifetime. The chance that a fan-fiction author decides to shift style and write an experimental novel with a new voice is slim.

The Twilight example and the J. K. Rowling/Galbraith example demonstrate two sides of the question of how genre affects writing style. Rowling changed genres, yet her writing style was still distinct. Fan-fiction authors write the same exact genre, yet their voices remain quite distinct from one another.

Mosteller and Wallace would likely not be surprised by the success of their model on Twilight because their first test of the case was also on the study of two writers with similar backgrounds, writing in the same series. They postulated that because writers use a consistent voice, it was possible to tell them apart.

All I did here was replicate the simple equations to test their theory on novelists. Ninety-nine times out of 100 the two statisticians were right: Within the prose of every writer, whether obvious to the reader or not, there *is* an underlying fingerprint setting them apart from all other authors who anyone has ever read.

Chapter **4**

Write by Example

When famous writers give advice, people listen. The logic is simple: These are the few people who have made a living, who have achieved *renown*, by stringing together a few dozen letters and symbols in the right order. Most of us haven't figured out how to make a dollar out of our words. So if one of the success stories is willing to spill the secret, it's time to pay attention.

Before taking anyone's advice, though, there are a couple of simple considerations worth making:

1. Does the person giving advice actually *follow* their own advice?
2. Does anyone *else* who has succeeded follow that same advice?

If a successful writer says that some aspect of writing is essential but doesn't follow her own rule—or if she's the only writer who follows the rule while a flock of other successful writers doesn't—then maybe that rule isn't so essential. On the other hand, if everyone we look at does follow a certain rule, then we've found a real secret to writing at its best.

With all this in mind, I decided to test a range of tips offered up by famous authors. This chapter picks apart examples ranging from Strunk and White's advice in their famous *The Elements of Style* to Chuck Palahniuk's proscription against "thought verbs" like *understand* and *realize*.

Exclamation Points

Let's start with something simple. In his book *10 Rules of Writing* Elmore Leonard offers a rule about exclamation points. He states, "You are allowed no more than two or three per 100,000 words of prose." A writing rule in the form of a ratio is a blessing for a statistician, so I ran with it.

Elmore Leonard was prolific. He wrote more than 40 novels in his career and 19 of his works have been adapted for the screen so far, including *3:10 to Yuma* and *Jackie Brown*. With a long and successful career he had time to fine-tune his preferences, down to the dots at the end of his sentences.

Leonard's 45 novels totaled 3.4 million words. If he were to follow his own advice he should have been allowed only 102 exclamation points his entire career. In practice, he used 1,651. That's *16 times* as many as he recommends! (!!!!!!!!!!!!!!!!)

But before you start thinking that Elmore Leonard was a secret exclamation point fanatic, consider the chart opposite. I counted the exclamation point usage of fifty authors, many considered the greatest writers in modern literature and many huge commercial successes, in their 580-plus books (the sample was the entirety of each author's novels unless otherwise noted), and the results ranged wildly.

Use of Exclamation Points
per 100,000 Words

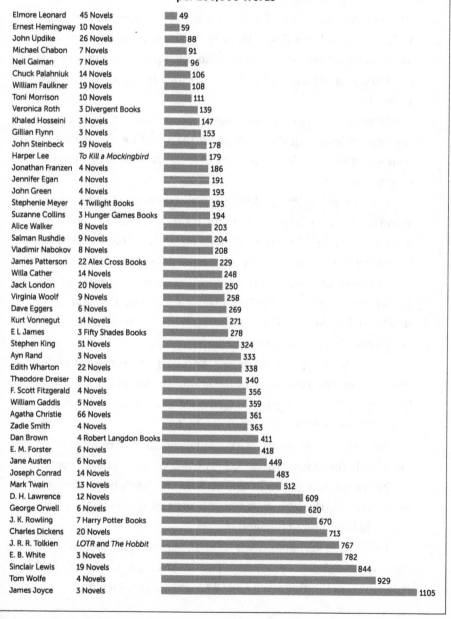

Elmore Leonard	45 Novels	49
Ernest Hemingway	10 Novels	59
John Updike	26 Novels	88
Michael Chabon	7 Novels	91
Neil Gaiman	7 Novels	96
Chuck Palahniuk	14 Novels	106
William Faulkner	19 Novels	108
Toni Morrison	10 Novels	111
Veronica Roth	3 Divergent Books	139
Khaled Hosseini	3 Novels	147
Gillian Flynn	3 Novels	153
John Steinbeck	19 Novels	178
Harper Lee	*To Kill a Mockingbird*	179
Jonathan Franzen	4 Novels	186
Jennifer Egan	4 Novels	191
John Green	4 Novels	193
Stephenie Meyer	4 Twilight Books	193
Suzanne Collins	3 Hunger Games Books	194
Alice Walker	8 Novels	203
Salman Rushdie	9 Novels	204
Vladimir Nabokov	8 Novels	208
James Patterson	22 Alex Cross Books	229
Willa Cather	14 Novels	248
Jack London	20 Novels	250
Virginia Woolf	9 Novels	258
Dave Eggers	6 Novels	269
Kurt Vonnegut	14 Novels	271
E L James	3 Fifty Shades Books	278
Stephen King	51 Novels	324
Ayn Rand	3 Novels	333
Edith Wharton	22 Novels	338
Theodore Dreiser	8 Novels	340
F. Scott Fitzgerald	4 Novels	356
William Gaddis	5 Novels	359
Agatha Christie	66 Novels	361
Zadie Smith	4 Novels	363
Dan Brown	4 Robert Langdon Books	411
E. M. Forster	6 Novels	418
Jane Austen	6 Novels	449
Joseph Conrad	14 Novels	483
Mark Twain	13 Novels	512
D. H. Lawrence	12 Novels	609
George Orwell	6 Novels	620
J. K. Rowling	7 Harry Potter Books	670
Charles Dickens	20 Novels	713
J. R. R. Tolkien	*LOTR* and *The Hobbit*	767
E. B. White	3 Novels	782
Sinclair Lewis	19 Novels	844
Tom Wolfe	4 Novels	929
James Joyce	3 Novels	1105

Elmore Leonard did not live up to his own advice in absolute terms, but relative to other authors he used the exclamation point with great rarity. Leonard did offer one caveat to his rule: "If you have the knack of playing with exclaimers the way Tom Wolfe does, you can throw them in by the handful." His assessment was dead-on. Wolfe used exclamation points at a rate of 929 per 100,000, a higher number than everyone in the sample except James Joyce.

When it comes to his own usage, perhaps Leonard was just not the best estimator. It wouldn't be surprising if he chose 100,000 words as it's a nice clean number which happens to amount to the length of a longish novel.

Another possibility is that Leonard did not start paying close attention to his exclamation point totals until he sat down to gather writing tips. The strict limit he offered to others could have been a rule he began to aspire to *after* he delivered it.

Consider the table on the next page showing his use of the exclamation point in each of his 45 novels. Leonard first stated his rule in 2001 in the *New York Times*. The bars in lighter gray are the books released after the rule surfaced in 2001.

At the beginning of his career, Leonard was throwing exclamation points into his books by the hundreds. All told, in the books he published before 2001, Leonard used exclamation marks at a rate of 57 per 100,000. After 2001, it was 10.

Leonard's eight books with the lowest rate were all written after 2001. The one exception to his post-2001 ways was *A Coyote's in the House*, which was Leonard's sole children's book. Perhaps he thought a little more excitement was needed to keep the attention of his new young readers.

Of all 580-plus books in my sample, only two can claim to have obeyed the strict "no more than two or three per 100,000 words" rule. One was Hemingway's *The Old Man and the Sea* with a single

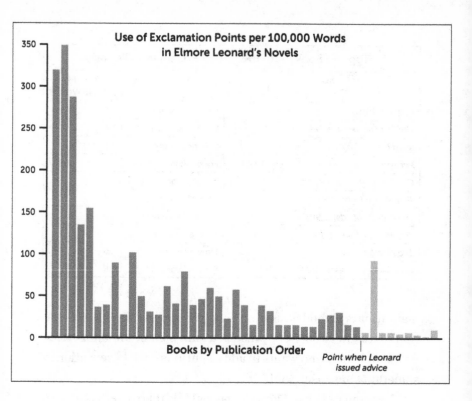

Use of Exclamation Points per 100,000 Words
in Elmore Leonard's Novels

Books by Publication Order

Point when Leonard
issued advice

"Now!" The other was Leonard's second-to-last novel, *Djibouti*, published in 2010, which also used just one exclamation point. On the following page are the ten books in my fifty-author sample with the fewest exclamation points. Notice the years on Leonard's books.

Leonard did not explain why he despises the exclamation point. But he's hardly alone in his opinion. Many style guides caution against its overuse, explaining that a flood of exclamation points will water down *all* moments of a text, and that the exclamation point should be reserved for those rare actions and descriptions that warrant extra attention. Fowler's *Dictionary of Modern English Usage* advises, "Except in poetry the excla-

Top Ten Books with the Fewest Exclamation Points		
BOOK	AUTHOR	! RATE PER 100,000 WORDS
Djibouti (2010)	Elmore Leonard	1.3
The Old Man and the Sea (1952)	Ernest Hemingway	3.6
Road Dogs (2009)	Elmore Leonard	4.1
Across the River and Into the Trees (1950)	Ernest Hemingway	4.3
Comfort to the Enemy (2006)	Elmore Leonard	5.4
True at First Light (1999)	Ernest Hemingway	5.9
The Garden of Eden (1986)	Ernest Hemingway	6.0
Tishomingo Blues (2002)	Elmore Leonard	6.2
The Hot Kid (2005)	Elmore Leonard	6.6
Up in Honey's Room (2007)	Elmore Leonard	6.6

mation mark should be used sparingly. Excessive use of exclamation marks in expository prose is a sure sign of an unpractised writer or of one who wants to add a spurious dash of sensation to something unsensational."

I wanted to know if Fowler was right: Is there a difference between "practiced writers" and "unpractised writers" in their punctuation usage? I had to generalize (I'm sorry fan-fic writers), but for my "unpractised" group, I downloaded all stories of at least 60,000 words posted on FanFiction.net since 2010 in the 25 most popular book universes. In total, the 9,000-plus works combined for more than a billion words. Then I compared their usage to my "practiced" group: the 100 most recent bestsellers and 100 most recent books to win literary awards.*

It turns out there is a drastic difference in the use of exclamation points. The median *New York Times* bestseller used 81 exclamation points per 100,000 words. Modern literary award winners

---

* This sample was created by sampling the most recent books to win the literary awards described in Chapter 2.

used 98 per 100,000 words. And fan-fiction authors used a median of 392 exclamation points per 100,000 words—about four times as many as our "practiced" group.

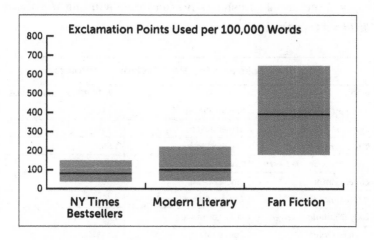

Too many exclamation points can come off as a sign of a writer relying wholly on a device to make their dialogue more exciting. Consider the six straight lines of dialogue below:

> "What's the matter with you!"
> "Let me go!"
> "You think I came to see you!"
> "Take your hands off of me!"
> He shook her violently. "You think I came for you!"
> "I don't care why you came!"

That's not from a fan-fiction story but a passage from Leonard's second novel, *The Law at Randado*, where he neared 350 exclamation points per 100,000 words. It was Leonard's second book and the author would never return to that level of forced excitement again.

It's important to note, though, that the question of any word or punctuation frequency is not completely clear-cut. There are plenty of classics, written by what Fowler would certainly deem "practised writers," that use exclamation points at an exceptional pace. A rate of 2,000 per 100,000 means around six per page. Of all books in the sample, Rushdie's Booker Prize–winning *Midnight's Children* tops the list.

Top Ten Books with the Most Exclamation Points		
BOOK	AUTHOR	! RATE PER 100,00 WORDS
Midnight's Children	Salman Rushdie	2,131
Finnegans Wake	James Joyce	2,102
The Chimes	Charles Dickens	1,860
The Cricket on the Hearth	Charles Dickens	1,793
Elmer Gantry	Sinclair Lewis	1,352
A Christmas Carol	Charles Dickens	1,351
Lady Chatterley's Lover	D. H. Lawrence	1,348
Back to Blood	Tom Wolfe	1,341
Dodsworth	Sinclair Lewis	1,274
Babbitt	Sinclair Lewis	1,144

Nonetheless, it's clear that amateur fan-fiction writers use the punctuation mark way more than professional writers. On the whole, I'd say that Leonard's advice holds. He follows it himself and it's common enough throughout successful writing to be worth paying heed to. Is that extra exclamation point really necessary? Or is it, as Fowler says, just "a spurious dash of sensation"?

Suddenly!

Just as the exclamation point can be a key or crutch, there are certain words that can make or break a great moment. They can provide just the right touch of drama or shatter a scene altogether. For a perfect example, let's look at the sixth rule of Elmore Leonard's *10 Rules of Writing*: Never use the word *suddenly*.

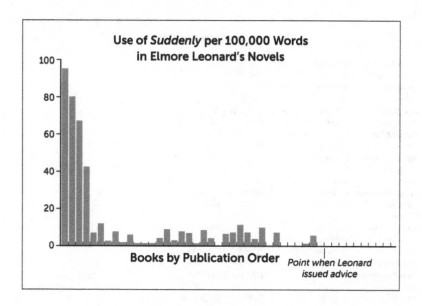

Use of *Suddenly* per 100,000 Words in Elmore Leonard's Novels

100
80
60
40
20
0

Books by Publication Order   *Point when Leonard issued advice*

Just about everyone uses the word *suddenly*, and like the exclamation point, Leonard used it *a lot* during his early career before giving it up nearly altogether.

In the nine novels he published after 2001 he never used the word *suddenly*. When Leonard said "never," he meant it.

Despite Leonard's early use of the word, his overall frequency of *suddenly* still ranks third lowest—behind Chuck Palahniuk and Jane Austen. On the other end of the spectrum, with career usages of about 70 per 100,000, are J. R. R. Tolkien, Joseph Conrad, and F. Scott Fitzgerald.

In the sample of 580-plus books, there are only 26 books that never use the word *suddenly*. Fifteen of these were written by Leonard. The remaining 11 were by Chuck Palahniuk, Mark Twain, or Stephen King.

Unlike exclamation points, *suddenly* doesn't have an enormous difference between fan-fiction writers and professionals. The same sample of fan fiction had a median *suddenly* rate of 22 per

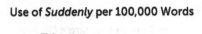

Use of *Suddenly* per 100,000 Words

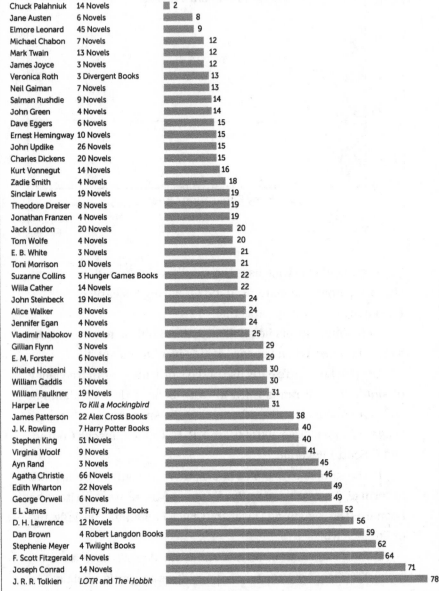

Chuck Palahniuk	14 Novels	2
Jane Austen	6 Novels	8
Elmore Leonard	45 Novels	9
Michael Chabon	7 Novels	12
Mark Twain	13 Novels	12
James Joyce	3 Novels	12
Veronica Roth	3 Divergent Books	13
Neil Gaiman	7 Novels	13
Salman Rushdie	9 Novels	14
John Green	4 Novels	14
Dave Eggers	6 Novels	15
Ernest Hemingway	10 Novels	15
John Updike	26 Novels	15
Charles Dickens	20 Novels	15
Kurt Vonnegut	14 Novels	16
Zadie Smith	4 Novels	18
Sinclair Lewis	19 Novels	19
Theodore Dreiser	8 Novels	19
Jonathan Franzen	4 Novels	19
Jack London	20 Novels	20
Tom Wolfe	4 Novels	20
E. B. White	3 Novels	21
Toni Morrison	10 Novels	21
Suzanne Collins	3 Hunger Games Books	22
Willa Cather	14 Novels	22
John Steinbeck	19 Novels	24
Alice Walker	8 Novels	24
Jennifer Egan	4 Novels	24
Vladimir Nabokov	8 Novels	25
Gillian Flynn	3 Novels	29
E. M. Forster	6 Novels	29
Khaled Hosseini	3 Novels	30
William Gaddis	5 Novels	30
William Faulkner	19 Novels	31
Harper Lee	*To Kill a Mockingbird*	31
James Patterson	22 Alex Cross Books	38
J. K. Rowling	7 Harry Potter Books	40
Stephen King	51 Novels	40
Virginia Woolf	9 Novels	41
Ayn Rand	3 Novels	45
Agatha Christie	66 Novels	46
Edith Wharton	22 Novels	49
George Orwell	6 Novels	49
E L James	3 Fifty Shades Books	52
D. H. Lawrence	12 Novels	56
Dan Brown	4 Robert Langdon Books	59
Stephenie Meyer	4 Twilight Books	62
F. Scott Fitzgerald	4 Novels	64
Joseph Conrad	14 Novels	71
J. R. R. Tolkien	*LOTR* and *The Hobbit*	78

100,000 words, while the samples of bestseller and award-winning novels written in the same time frame come in at 16 and 19, respectively. The usage is slightly lower in the pros, but too similar to be universalized into a rule.

Sometimes, and with some writers, like super-user J. R. R. Tolkien, its usage can be too obvious. (Take this example: *The valley seemed to stretch on endlessly. Suddenly Frodo saw a hopeful sign.*) But this is one case where Leonard's "never" seems over the top. Finely tuned bestselling books use *suddenly* just as much as unedited fan-fiction. And while Leonard indeed practiced what he preached, he's one of the *only* authors to write entire books without it. His hardcore stance, as seen through the wide range of other authors using it, comes off as extreme. Perhaps the better piece of advice would not be to suddenly stop altogether—but to use the word in moderation.

Thought Verbs

Chuck Palahniuk's writing is distinct. We just saw him land on the extreme ends of both exclamation-point and *suddenly* usage, and in Chapter 1 we saw his avoidance of -ly adverbs. He has explained many of his own theories on writing, and one of his most interesting recommendations is to avoid the use of "thought verbs."

As the author explains in a 2003 essay, "Instead of characters *knowing* anything, you must now present the details that allow the reader to know them. Instead of a character *wanting* something, you must now describe the thing so that the reader wants it." It's an idea very similar to other writers' proscriptions against exclamation points, *suddenly*, and adverbs in general: Don't rely on a single device to create an atmosphere that you could instead create in context, with a chorus of other words and punctuation marks working in concert. With his denouncement of "thought verbs,"

Use of Thought Verbs per 10,000 Words

James Joyce	3 Novels	56
J. R .R. Tolkien	*LOTR* and *The Hobbit*	60
Chuck Palahniuk	14 Novels	64
Vladimir Nabokov	8 Novels	64
Jack London	20 Novels	74
Mark Twain	13 Novels	75
Charles Dickens	20 Novels	77
Michael Chabon	7 Novels	78
Willa Cather	14 Novels	80
Salman Rushdie	9 Novels	85
Joseph Conrad	14 Novels	85
Virginia Woolf	9 Novels	85
George Orwell	6 Novels	87
Dan Brown	4 Robert Langdon Books	88
Kurt Vonnegut	14 Novels	89
John Updike	26 Novels	89
E. B. White	3 Novels	92
William Faulkner	19 Novels	93
Khaled Hosseini	3 Novels	94
Neil Gaiman	7 Novels	94
J. K. Rowling	7 Harry Potter Books	96
Edith Wharton	22 Novels	97
Sinclair Lewis	19 Novels	98
Tom Wolfe	4 Novels	98
Jane Austen	6 Novels	101
Harper Lee	*To Kill a Mockingbird*	101
Jonathan Franzen	4 Novels	103
Jennifer Egan	4 Novels	104
William Gaddis	5 Novels	104
F. Scott Fitzgerald	4 Novels	105
D. H. Lawrence	12 Novels	111
Suzanne Collins	3 Hunger Games Books	113
James Patterson	22 Alex Cross Books	114
Stephen King	51 Novels	114
Zadie Smith	4 Novels	114
E. M. Forster	6 Novels	117
John Steinbeck	19 Novels	117
Theodore Dreiser	8 Novels	119
Toni Morrison	10 Novels	126
Veronica Roth	3 Divergent Books	128
Stephenie Meyer	4 Twilight Books	128
Ernest Hemingway	10 Novels	132
John Green	4 Novels	134
Dave Eggers	6 Novels	135
E L James	3 Fifty Shades Books	140
Gillian Flynn	3 Novels	141
Ayn Rand	3 Novels	144
Agatha Christie	66 Novels	144
Alice Walker	8 Novels	145
Elmore Leonard	45 Novels	150

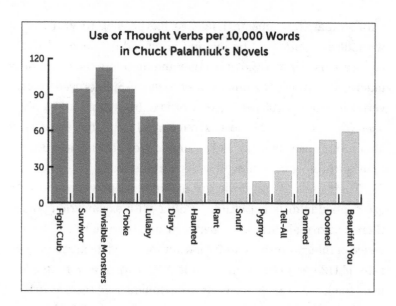

Use of Thought Verbs per 10,000 Words in Chuck Palahniuk's Novels

Palahniuk is cutting to the core of another oft-repeated piece of writing advice: "Show; don't tell."

He identifies thought verbs as "Thinks, Knows, Understands, Realizes, Believes, Wants, Remembers, Imagines, Desires, and a hundred others you love to use." He singles out *loves* and *hates* later on in his essay. And for our purposes, we'll limit the search to these 11 words (and their other tenses).

Like Leonard, Palahniuk lives by his own advice and ranks near the top of the list for fewest thought verbs.

And also like Leonard, it's unclear what came first: his radical desire to eliminate thought verbs from his writing or his advice to do so. His eight books with the lowest rate of thought verbs are the eight books he wrote after his 2003 essay. Palahniuk's usage dropped almost in half from an average of 88 to 45 thought verbs per 10,000 words. In the graph above, the light shade indicates books published after 2003.

Palahniuk's advice is different from Leonard's because it deals with ordinary fundamental language. It is possible to write a book without *suddenly* or ! by making minor adjustments. Thought verbs are harder to avoid altogether and therefore the variation is much less extreme. Romance writer Nicholas Sparks used thought verbs at a rate of 200 per 10,000 words in his major hit *The Notebook*. That's close to the most extreme rate I could find, and it's just about four times as many as Palahniuk uses post-advice. But did Sparks's profusion of thought verbs hurt his work?

There are a few takeaways from looking at how thought verbs are applied in different writing. Unlike with exclamation points, there is no real difference between amateur and professional usage of thought verbs. Fan-fiction authors used them at a median rate of 112 per 10,000 words, while *New York Times* bestsellers used them at 113 and modern award winners used them at 104.

For modern writing, in fact, a clear indicator of thought verb usage is genre. Of the 100 most recent *New York Times* bestsellers, 13 were romance* and they came in at an average of 145 per 10,000. Many popular romance novelists come in on the extreme end as well—E L James at 140, Nora Roberts at 143, and Nicholas Sparks averages 168.

As someone who writes with twisted characters and dark humor, there's no way Palahniuk is a fan of the romance genre. In 2011 he wrote a short story for *Playboy* called "Romance" in which a man believes he is dating a high-functioning alcoholic but it turns out he is dating a high-functioning woman with a mental disability.

Are You There God? It's Me, Margaret by Judy Blume is a classic coming-of-age book for young girls in which Margaret deals with her parents and puberty. It's told in a diary where each entry starts with the same form as its title. Palahniuk's *Damned* is a

---

* I used Goodreads genre tags for classifying the books. I counted books categorized as "romance" but did not count genre-spanning books, such as "fantasy-romance."

parody of the book with each entry, penned by a thirteen-year-old girl in hell, starting off "Are you there, Satan?" While Blume's book uses 182 thought verbs per 10,000, *Damned* uses just 46.

Palahniuk's disdain for heartwarming stories speaks to the type of stories the *Fight Club* author is aiming to tell, but not as much about all stories anyone may aspire to write. In this sense, Palahniuk's advice may only be applicable for a very small niche. For writers aiming to tell stories of protagonists transgressing the norms of society, staying away from direct mention of emotion might be best. For everyone else, spelling out the thoughts and dreams of your characters to the reader is far from the end of the world.

Qualified Advice

In 1920, Cornell professor William Strunk, Jr. published *The Elements of Style*. It was an obscure guidebook until 39 years later, when *Charlotte's Web* author E. B. White updated it with his own advice. Since then it has earned the shorthand name of simply *Strunk and White*, and is considered by many the standard introduction to the craft of writing.

Much of the advice is familiar if you've read any book on writing (including the previous chapters of this book). It warns against emphasizing "simple statements by using a mark of exclamation" and encourages the reader to "write with nouns and verbs, not adjectives and adverbs." The book also declares it is best to write with force and certainty. In the revised edition White writes: "Consciously or unconsciously, the reader is dissatisfied with being told only what is not; the reader wishes to be told what is. Hence, as a rule, it is better to express even a negative in positive form."

The examples he gives are of using "dishonest" instead of "not honest" and the sentence "He usually came late" instead of "He was not very often on time." So let's look at the word *not*. How does E. B. White do in living up to his own advice?

Use of *Not* per 10,000 Words

James Joyce	3 Novels	52
Dan Brown	4 Robert Langdon Books	61
Michael Chabon	7 Novels	66
Chuck Palahniuk	14 Novels	67
Virginia Woolf	9 Novels	68
Vladimir Nabokov	8 Novels	71
E. B. White	3 Novels	75
Khaled Hosseini	3 Novels	76
Salman Rushdie	9 Novels	77
Kurt Vonnegut	14 Novels	77
F. Scott Fitzgerald	4 Novels	77
Jack London	20 Novels	80
Jennifer Egan	4 Novels	80
James Patterson	22 Alex Cross Books	82
Willa Cather	14 Novels	84
Charles Dickens	20 Novels	84
John Updike	26 Novels	85
Edith Wharton	22 Novels	89
Joseph Conrad	14 Novels	93
Stephen King	51 Novels	93
Tom Wolfe	4 Novels	94
Neil Gaiman	7 Novels	95
Alice Walker	8 Novels	95
J. R. R. Tolkien	*LOTR* and *The Hobbit*	96
Jonathan Franzen	4 Novels	97
George Orwell	6 Novels	97
J. K. Rowling	7 Harry Potter Books	98
Sinclair Lewis	19 Novels	99
Mark Twain	13 Novels	100
D. H. Lawrence	12 Novels	101
Dave Eggers	6 Novels	102
E L James	3 Fifty Shades Books	103
Gillian Flynn	3 Novels	104
John Steinbeck	19 Novels	106
Zadie Smith	4 Novels	108
John Green	4 Novels	111
Toni Morrison	10 Novels	112
Suzanne Collins	3 Hunger Games Books	113
Harper Lee	*To Kill a Mockingbird*	113
William Gaddis	5 Novels	114
Elmore Leonard	45 Novels	116
Theodore Dreiser	8 Novels	124
Ernest Hemingway	10 Novels	125
Jane Austen	6 Novels	126
William Faulkner	19 Novels	131
Agatha Christie	66 Novels	131
Stephenie Meyer	4 Twilight Books	131
E. M. Forster	6 Novels	137
Veronica Roth	3 Divergent Books	146
Ayn Rand	3 Novels	151

He's up near the top, seventh on our list of fifty authors. (For those wondering if White's books having low usage here has to do with the fact his novels were children's books, I don't think this is the case. I checked other children's authors, such as Roald Dahl, who did not compare to White.)

The word *not* is so innocuous and fundamental that it seems hard to imagine a huge difference between practiced writers and unpracticed. In fact, when going through samples of current writing the variation was slim to none. In the modern literary sample it's used at a rate of 88 per 10,000 words, 100 times in modern bestsellers, and 103 times in fan fiction. *Not* shows some variation, but the difference is so minor that it's hard to see *not* as a great indicator of quality or professionalism for current writing.

But other advice that *Strunk and White* touches on does rely on discrete, extraneous words. The book is full of recommendations to build a simple, straightforward writing style. One of the key items, added by White to the 1959 edition, advises writers: "Avoid the use of qualifiers." White picks out several words in particular: "*Rather, very, little, pretty*—these are the leeches that infest the pond of prose, sucking the blood of words."

Now, one of White's three books is *Stuart Little*, so out of fairness we'll strip *little* from the count. Looking at just *rather, very*, and *pretty*, what White describes as leeches on prose, how does the *Charlotte's Web* author measure up?

Unlike the other tips I showed earlier, White does not follow his own advice when compared to others. And though there is variation from one author to another, these differences are not apparent in any large quantity across different types of modern writing. Modern award winners come in at 108 per 100,000 words, 118 for bestsellers, and 103 for fan fiction.

This all looks like a loss for White: He used the qualifiers at a high rate of 220 per 100,000 and the pros use qualifiers just as often as the amateurs.

But what if, in the big picture, White has won? What if his advice is being followed by *all* writers and *all* speakers?

The edition of *The Elements of Style* in which White elaborated on the use of qualifiers came out in 1959. In the 1960s the median bestseller used 152 per 100,000. Qualifier use had declined more than 20% by 2010's bestsellers.

If we expand the time frame the difference is clearer and more extreme. Using the subset of twentieth-century classics (as described in more detail in Chapter 2), we can see that use has declined throughout the past century. Classics of the earliest part of the twentieth century, like *Heart of Darkness* and *My Ántonia*, had double the qualifier usage of newer classics like *Beloved* and *Possession*.

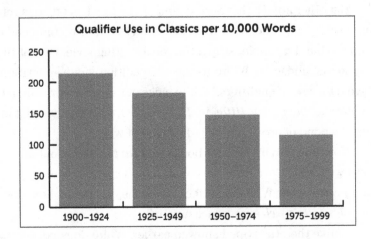

And if we extend even further back the trend continues. I looked at all top downloads from Project Gutenberg's ebook collection. These are classic books such as *Pride and Prejudice*, *The Strange Case of Dr. Jekyll and Mr. Hyde*, and *The Importance of Being Earnest*. Of the books written between 1850 and 1899 the median qualifier usage was 260, and between 1800 and 1849 qualifier usage measures in at 297. That's almost three times as much as what you could expect to find in any literary or popular book today; the current average is just over 100.

In the long game, E. B. White has won, even if he wasn't the paragon of the movement. Qualifier use has been declining for centuries. Much of the decline has to do with the decline of the word *very* which accounts for about 75% of the three qualifiers White singled out.

Jane Austen is one of the English language's most celebrated authors but her use of words like *very* is off the charts. Her book *Pride and Prejudice* has a rate of 396 uses of *very* per 100,000 words. Consider this excerpt:

> They were not welcomed home **very** cordially by their mother. Mrs. Bennet wondered at their coming, and thought them **very** wrong to give so much trouble, and was sure Jane would have caught cold again. But their father, though **very** laconic in his expressions of pleasure, was really glad to see them; he had felt their importance in the family circle.

In Austen's writing *very* is common, peppered throughout paragraphs at a rate that might seem jarring today. But much of that has to do with the era. In the early 1800s her use was high, but not atypical. Mary Shelley's *Frankenstein* and Charles Dickens's *Oliver Twist* weigh in, respectively, at 103 and 304 uses of *very* per 100,000 words. If you compare these to modern literary works, such as the 2017 Pulitzer Prize–winning *The Underground Railroad*, which has just 30 uses of *very* per 100,000 words, it becomes clear that acceptable usage of the word has changed over time.

There's no way to say how much this has to do with people editing out unneeded words, as Strunk and White advise, and how much has to do with a longer-term change in the way people talk. I think at least part of it, though, does have to do with people being more attentive to the granular elements of their writing. Coming up through an average American public school, I remember being told to delete the *very*s from my papers even when more nuanced writing tips were nowhere to be found. The tip even made its way into *Dead Poets Society* (the book was inspired by the movie, not the other way around), with a rationale that sounds very similar to the argument for avoiding adverbs (well, at least the first part does . . .).

> So avoid using the word "very" because it's lazy. A man is not very tired, he is exhausted. Don't use very sad, use morose. Language was invented for one reason, boys—to woo women—and, in that endeavor, laziness will not do. It also won't do in your essays.

The irony, of course, is that given the decline of *very*, the students in the Dead Poets Society would have been more likely to read the word *very* in the works of the deceased writers they admired than anything they were already writing.*

Throughout this chapter, I've dug through a handful of writing tips from acclaimed authors. Writers are generally good at following through with their own advice. But the tricky part for us is that they often give advice on matters that contrast with the style of other successful writers. Are these writing tips universal means to improvement? Or are they nothing more than a novelist's pet peeves? In some cases the evidence is clearer than others. I would trim the use of exclamation marks and the word *very* based on the data alone, even if the advice had never been dispensed.

---

* Google Books Ngrams also shows a decline in the word *very* over time. From 1900 to 2000 it dropped about 60%. However, Google Books includes books such as *Very Long Baseline Interferometer* and *My Very First Mother Goose* and there is no standardization of genre over time. Therefore, Google Books Ngrams alone may not be representative of usage of words in popular and literary fiction.

In other cases, the data shows that writers advise things they are doing on a unique level. These are fascinating habits to consider, but not necessarily details that everyone should incorporate into their own writing. Here, the more important lesson may lie in the pure attention to detail that these rules inspire. It may not be the avoidance of thought verbs that makes Palahniuk a great writer, but rather the fact that he's scrutinizing the effects of even such seemingly straightforward words in his work. It's by noting the role of each word and punctuation mark that the greats are able to hone their writing. Ultimately, this may be what gives the words that remain on the page—in a classic or a bestseller—their power to move us.

Chapter 5

Guiltier Pleasures

One day I will find the right words, and they will be simple.
—JACK KEROUAC, *THE DHARMA BUMS*

If you have ever read a Dr. Seuss book, you may be familiar with words like *fizza-ma-wizza-ma-dill*, *fiffer-feffer-feff*, and *truffula*.

You may also be familiar with these: *a, will, the*.

Besides made-up words and rhymes, Dr. Seuss's biggest trademark is the simplicity of his writing. Even compared to other children's authors, Dr. Seuss pushed the limits. We can partly thank his Houghton Mifflin editor, William Spaulding, who after a string of successes presented Seuss with a list of just a few hundred simple words in the mid-1950s. Seuss had already published *Horton Hears a Who!*, *And to Think That I Saw It on Mulberry Street*, and *If I Ran the Zoo*. But, as detailed in the *New Yorker* article "Cat People," Spaulding wanted Seuss to go after an even younger audience: "Write me a story that first graders can't put down!"

Seuss would later describe how he struggled with Spaulding's challenge:

He sent me a list of about three hundred words and told me to make a book out of them. At first I thought it was impossible and ridiculous, and I was about to get out of the whole thing; then decided to look at the list one more time and to use the first two words that rhymed as the title of the book—*cat* and *hat* were the ones my eyes lighted on. I worked on the book for nine months—throwing it across the room and letting it hang for a while—but I finally got it done.

The result was *The Cat in the Hat*. It clocks in at 220 unique words, and to this day ranks as the second-most-selling book of Seuss's career. The one book ahead of it? It's *Green Eggs and Ham*, which uses just fifty words. All but one, *anywhere*, are one syllable.

Seuss's two most popular books are those in which he restricted himself the most: Simplicity brought success.

Of course, Seuss was not writing for a general audience—he was writing for children still learning to read. It would be impossible to write a book with just fifty words if it weren't covered in giant illustrations. And adult readers are looking for more than your average first-grader.

But what if there's more to this idea? Sure, the books we read and love as adults are more complex—but just how complex? Is there an ideal level if you're aiming to write the next number one bestseller? And where do the literary greats clock in?

The word lists that Dr. Seuss used when writing *The Cat and the Hat* and *Green Eggs and Ham* were created by a man named Rudolf Flesch. In his 1955 book *Why Johnny Can't Read*, Flesch argued that reading education in America was in dire need of reform; he introduced the nation to phonics and his word lists ended up inspiring just the kind of revolution he was hoping for.

Flesch then went on to create a mathematical formula—the Flesch-Kincaid Grade Level test—that was capable of measuring the simplicity or complexity of any text. The formula itself is simple—just a couple of fractions weighted and then added together.

$$0.39 \left(\frac{\text{total words}}{\text{total sentences}} \right) + 11.8 \left(\frac{\text{total syllables}}{\text{total words}} \right) - 15.59$$

The score, by Flesch's description, is the grade level required to understand the text. If a book is given a grade of 3, that means a third grader (and anyone older) could be expected to understand it.

The test works best when applied to large texts but it's easy to understand with short samples. Take the first sentence in George Washington's first State of the Union Address:

I embrace with great satisfaction the opportunity which now presents itself of congratulating you on the present favorable prospects of our public affairs.

At 43 syllables and 23 words, the sentence would be given a grade-level score of 15.

Compare to the first sentence in George W. Bush's last State of the Union Address:

Seven years have passed since I first stood before you at this rostrum.

It has 16 syllables, 13 words, and a grade-level score of 4.

The numbers 4 and 15 may seem arbitrary, but side by side it's easy to see why the first sentence is given the higher complexity score. The grade scores tend to average out over the course of a longer text, but short samples like these do show the limits of a metric like Flesch-Kincaid. It has its detractors, who criticize the formula for being too simple or not capturing context or fumbling the exact grade levels.

For instance, there are some unusual writers whose unique style breaks the simple scoring system. *Green Eggs and Ham* has a negative grade-level score, -1.3 to be exact. Consider the passage below:

> Not in a box.
> Not with a fox.
> Not in a house.
> Not with a mouse.
> I would not eat them here or there.

It has 24 words and 24 syllables spread out over 5 sentences, which yields a negative score.

On the other end of the spectrum is William Faulkner. In *The Sound and the Fury* he disregards punctuation, which leaves him with a "sentence" composed of over 1,400 words. It has a Flesch-Kincaid Grade Level score of 551.

But these are the outliers, the texts that pose the biggest challenge. As a relative measure, Flesch-Kincaid works well, averaging out the irregular sentences over the length of a full book. Even *The Sound and the Fury* as a whole has a grade-level score of 20. Most books meant for a general audience, not by Faulkner and not by Seuss, will fall within the fourth to eleventh grade. Every *New York Times* number one bestseller since 1960 falls in this seven-grade-level spread.* Ultimately, Flesch-Kincaid's simplicity is an advantage, allowing us to compare huge swathes of text across genres or generations.

If you follow U.S. politics you might see the Flesch-Kincaid formula pop up once a year when it's time for the State of the Union. It has become a popular pastime to evaluate the complexity of these speeches, for doing so reveals an undeniable trend. When comparing all of the State of the Union addresses from America's founding

---

* About 15% of books from the earlier decades are missing from the sample because they are not available in digital form.

to the present, the Flesch-Kincaid test shows a steady decline in the sophistication of our political speech.

If you're being optimistic, politics is reaching a wider audience. If you're being cynical, politics is getting stupider by the decade.

In an article in the *Guardian*, titled perfectly "The state of our union is . . . dumber," the authors used Flesch-Kincaid to determine that the annual presidential State of the Union Address has gone from an eighteenth-grade level pre-1900, to a twelfth grade level in the 1900s, and it's now sunk below a tenth-grade level in the 2000s.

The role of the State of the Union has changed over the years. After all, when Washington was addressing Congress in 1790 it was meant as an actual address* to Congress. The event has transformed into a national radio and television spectacle, making it important to reach every corner of America, regardless of age and education.

It might be one thing for the State of the Union to drift downward, but what about the world of books? Do we see any patterns when we look at the state of the American novel over time? Is the state of our fiction . . . dumber?

To find out, I collected every digitized number one *New York Times* bestseller since 1960† and ran the Flesch-Kincaid test on all 563 of them.‡

---

* Up until the creation of radio and television the State of the Union was often a written document sent to Congress.

† The 2010s only cover 2010–2014.

‡ About 15% of the books from the first few decades were not available in electronic form and were not included in this analysis. However, even if they all were written at an extremely low level it would not be enough to move the median of these decades below modern levels.

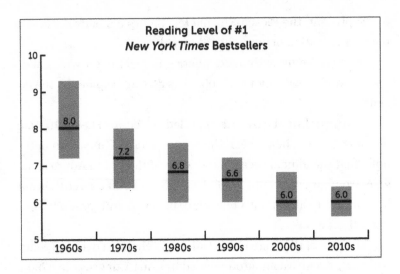

Reading Level of #1
New York Times **Bestsellers**

The overall trend, in just the last fifty-plus years, shows the same downward slope. The bestseller list is full of much simpler fiction. If you pick books by checking out what is trending on the list, chances are you are going to be picking up books of less sophistication today than you would forty or fifty years ago.

The black bar represents the reading level of the median book in each decade. The shaded region represents the middle 50% of all books. In the 1960s the median book had a grade level of 8.0, with the middle half of all books coming in between 7.2 to 9.3. While 7.2 could be considered low fifty years ago, in 2014 there were 37 bestsellers, and 36 had a grade level of 7.2 or below. The floor for simplicity has become the ceiling. The number one bestseller with the highest reading level in 2014 was Daniel Silva's *The Heist*, which had a score of 8.0. Out of all 37 books it was the one book that had a score that half a century ago would have been typical.

On the upper end, James Michener's 1988 novel *Alaska* had a grade-level score of 11.1, making it the number one bestseller since 1960 with the highest reading level. Twenty-five books since 1960 have had a grade level of 9 or higher. But just two of these were written *after* 2000.

On the lower end, eight books tie for the lowest score of 4.4. All eight of these books were written after 2000, all by one of three high-volume writers: James Patterson, Janet Evanovich, and Nora Roberts.

These ultrapopular bestsellers with low reading levels are a recent phenomenon. Twenty-eight of the number one bestsellers since 1960 that I collected had a grade-level score below 5. Just two of these were written *before* 2000.

To see the trend before your eyes, below is a graph showing the percentage of books with a grade level greater than eight, the median score in the 1960s.

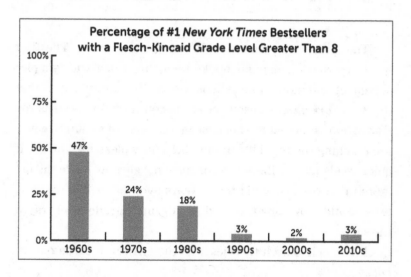

And on the next page is a graph showing the percentage of books with a grade level less than six. This is the median today.

The *New York Times* bestseller list holds a rarefied place in the book world. To have written a *New York Times* bestseller is to have made it. And for the general public, the list often serves as the public face of fiction, a guide to what's worth reading. Yet in the last fifty years, there is no way around it: The books that we're reading have become simpler and simpler.

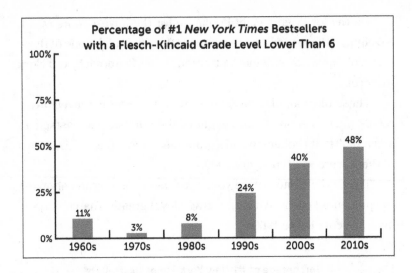

Percentage of #1 _New York Times_ Bestsellers with a Flesch-Kincaid Grade Level Lower Than 6

Decade	Percentage
1960s	11%
1970s	3%
1980s	8%
1990s	24%
2000s	40%
2010s	48%

There are two reasons this could be happening. The first would be that all popular books today are filled with simpler sentences and more monosyllabic words. The alternative is that the _New York Times_ bestseller list is getting "dumber"—as the _Guardian_ would put it—because more books of a "dumb" genre are reaching the top. I'll call this the "guilty pleasure" theory. If quick reads like thrillers and romance novels now make the list more often than they did thirty years ago, the median reading level would move down even if each genre's grade level stayed the same.

I've checked both theories, and the answer, it turns out, is: _both_.

There have always been "guilty pleasure" books on the list. In the 1960s it was _Valley of the Dolls_, in the 1970s, _The Exorcist_, in the 1980s, the books of the Bourne Trilogy, and in the 1990s The Lost World of the Jurassic Park series.

But without a doubt, there are more guilty pleasures on the list today than there used to be. In the 1960s a book would hold its top position on the list for many months at a time. Today, books jump up and down the chart much more rapidly. James Michener's *Hawaii* and Allen Drury's *Advise and Consent* were the only two books to reach number one bestseller status in 1960. In 2014 there were 37 that did, and the longest any one book claimed the top spot was four weeks (Grisham's *Gray Mountain*). Prize-winning literary novels like *The Corrections* and *The Goldfinch* make the number one spot on occasion, but today it's much more often dominated by commercial novels. This makes the contribution of the literary books less important to the median.

Looking at prizewinners rather than bestseller lists, we find that literary books haven't declined in reading level nearly as much. That being said, they are not as complicated in terms of sentence length and word length as you might think. Complicated themes don't always translate to complicated writing. The average for Pulitzer Prize winners in the 1960s was a 7.6 grade-level score, and in the 2000s a 7.1. In the years in between, the average was 7.4. There are many more outliers among the Pulitzer winners (Chabon's *The Amazing Adventures of Kavalier & Clay* scored a 10.0 while Alice Walker's *The Color Purple*, from the perspective of fourteen-year-old Celie, scores a 4.4), but there has not been a systematic shift over the years.

The growing presence of guilty pleasures is not the sole reason for the decline in bestseller reading levels, however. If we break down bestsellers by genre, we find that there has been a long-term shift within those guilty pleasures. Thrillers have become "dumber." Romance has become "dumber." There has been an across-the-board "dumbification" of popular fiction.

In the following graph I have plotted the 25 authors with the most number one bestsellers since 1960. All of these writers have had at least seven number one hits in their career, and just about all of them are writing for a broad audience: suspense, mystery, romance, action, etc. They are shown by the average reading level of their books and the year in which their first number one best-seller was written.*

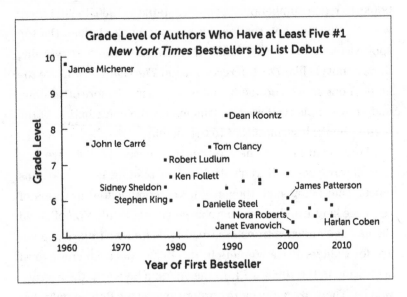

Robert Ludlum is known for thrillers like the Bourne Trilogy, which debuted in 1980, but he still wrote at a Flesch-Kincaid reading level of 7.2, not common in popular fiction today. Tom Clancy and Dean Koontz, who both got their starts in the 1980s, write at a level higher than any of the rising popular writers of the last twenty years. Your average John le Carré novel had a reading level higher than 36 of the 37 number one bestsellers from 2014. Danielle Steel ranks as a low outlier for her time period, but she still writes at a higher grade level than many of her even more modern counterparts.

There aren't just more guilty pleasures representing popular books. The pleasures have gotten guiltier.

---

* The cutoff to make the chart above was seven bestsellers by the end of 2014. There is a possible skew in the data due to the fact that in order for a writer starting to write in the 2000s, he or she had to write much more quickly (and therefore possibly more simply) to achieve seven bestsellers by the end of 2014.

Though it is the most prevalent, the Flesch-Kincaid test is just one of many tests of reading level. Most use sentence length as a large component. Today's bestsellers have much shorter sentences than the bestsellers of the past, a drop of 17 words per sentence in the 1960s to 12 in the 2000s. This means any of these similar tests will show similar declines.

One interesting alternative is the Dale-Chall readability formula. While it too uses sentence length, it has a separate component that factors in the number of "complex" words that appear in a text. In 1948 Edgar Dale and Jeanne Chall compiled a list of 763 words they did *not* consider complex. From this list it's possible to count the number of "complex" and the number of "not complex" words in a text.* The thought is that it's not just sentence length that can make a book hard to follow for young readers, but the number of words that are unfamiliar.

Since their original list, Dale and Chall have expanded the list to almost 3,000 words. Over 99% of the words in Dr. Seuss's *The Cat in the Hat* are considered "not complex." Seuss's only two exceptions are *thump(s)* and *plop*.

But 1% complex is unheard-of when it comes to novels. The closest any number one bestseller has come to this is Danielle Steel's 1993 *Star*, which uses a record-low 7% "complex" words. The book's first sentence is below with its one "complex" word bolded.

> The birds were already calling to each other in the early morning stillness of the Alexander Valley as the sun rose slowly over the hills, stretching golden fingers into a sky that **within** moments was almost purple.

On the other side we have Robert Ludlum's 1984 thriller, *The Aquitaine Progression*. Twenty-two percent of its words are con-

---

* Conjugates of verbs are accounted for. Proper nouns are discounted.

sidered "complex" by Dale and Chall, more than any other number one bestseller. Below are the first three sentences with words considered "complex" in bold.

> Geneva. City of sunlight and bright **reflections**. Of **billowing** white sails on the lake—**sturdy**, **irregular** buildings above, their **rippling images** on the water below.

Just as we see Flesh-Kincaid scores decline in recent decades, so do the "complex word" counts measured by Dale-Chall. Though the results are much less pronounced than the change in Flesch-Kincaid reading levels, there is a clear downward trend since 1960.

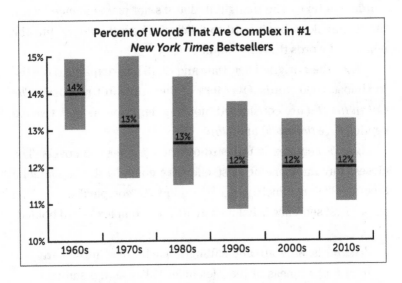

A bestseller once considered typical in complex word usage would now be on the high end of the spectrum. A 2% decrease is small in absolute terms but when you consider the small range a book may fall in, somewhere between 7 and 22%, a 2% fall is significant.

* * *

Where will the bestseller list be in ten or twenty years?

The *New York Times* bestseller list is looked up to in the book world. It's prestigious for authors and a guide for readers. By the changes the *New York Times* has made over the years, it's clear that they think about its composition. And while their exact methods are undisclosed, they've acknowledged that unlike other lists they weigh the sales of certain independent stores more heavily than bigger retailers, tending to give a fighting chance to smaller, more "literary" books versus the commercial, grocery-aisle thrillers. But if the *New York Times* is conscious of the trend in its bestseller list, they face a question: Should they ever step in to exclude certain authors or genres from the bestseller list for fiction?

It might sound outrageous at first for the *Times* to try to "shape" the list, but they have done it before. In 2000 a major change was made that excluded the Harry Potter books from the list. In the previous year, the number one spot was filled by a Harry Potter book on twenty separate weeks. The result was a new "Children's Book" list, which has since splintered even more into distinct "Young Adult," "Middle Grade," "Picture," and "Series" lists.

One obvious fix to the dominance of guilty pleasures would be to split up the fiction list, the marquee list of the *New York Times*, into one focused on literature and one focused on genre fiction. If the *New York Times* wanted to promote a diverse range of books, they could make the former list their most touted. A safe haven could at least please those serious readers who want to know what's popular in the world of books beyond pulp fiction. (Admittedly, the line between genre fiction and literary fiction would be difficult to draw, especially if publishers have a financial interest in gaining a certain categorization.)

A similar effort was already attempted by the *New York Times* editors. Though it did not alter their main fiction list, in 2007 a "trade" paperback section was unveiled, which was supposed to shine a light on a certain brand of fiction. Here was the description used by the *New York Times*'s own editors upon its release: "This issue also introduces a new bestseller list, devoted to trade paperback fiction. It gives more emphasis to literary novels and short-story collections...."

The counterpart to the "trade" paperback was the "mass-market" paperback. A book qualifies for the "mass-market" bestseller list based not on its genre or potential audience, but if it's printed within certain parameters (smaller pages, cheaper paper; often they are those pocket-sized editions you tend to see in grocery stores). And it just so happened that genre paperbacks tended to be printed as mass-market paperbacks. However, the market for these inexpensive books has since shrunk with the rise of ebooks, so more and more genre or commercial books are instead being published in the trade paperback format—that is, as higher quality paperbacks meant to be more lasting. As a result, the trade paperback list has not lived up to its initial selling point. As I wrote this chapter, the number one book on the trade paperback list throughout was the infamous erotica novel *Fifty Shades of Grey*. It was followed by *Fifty Shades Darker* and *Fifty Shades Freed*. All three have been on the list for more than 100 weeks. The rest of the list had some more literary fiction, but also works by authors Gillian Flynn, Nicholas Sparks, and James Patterson, who seem to defy the list's goal of giving "more emphasis to literary novels."

If the *New York Times* wants to accomplish that, they're going to again need to adjust their categories. Perhaps it's time to bite the bullet and try to define that elusive "literary" genre rather than base their decisions on the book's physical format. If they wish to hold on to their cultural perch, they'll likely need to change again as they have done before.

The broader question that I keep thinking back to, however, is: Should we worry at all about the overall reading level of the fiction bestseller list?

For this question, I say no. I have devoted an entire section to showing how bestsellers have become ... dumber. It would be easy for me to lump the *New York Times* reading level decline in with the rise of knee-jerk arguments that the country's intellect is at an all-time low.

But I don't think this is fair. Remember, the reading level is supposed to be a rough cutoff for who is excluded from a text. You don't have to be in sixth grade to read a book written at a sixth-grade level. Books with simpler texts can appeal to a wider audience.

Simple can be great. It includes more people. Writing doesn't need to be complicated to be considered either powerful or literary. The winner of the 2014 Pulitzer Prize for fiction, *The Goldfinch*, was also a number one bestseller and has a reasonable reading level of 7.2. While many classics do have high scores (*The Age of Innocence* at 10.4, *Oliver Twist* at 10.1, *The Satanic Verses* at 10.1), just as many have surprisingly low scores. *To Kill a Mockingbird* has a reading level of 5.9, *The Sun Also Rises* at 4.2, and *The Grapes of Wrath* measures down all the way at 4.1. All three of these books are revered by the literary community, but are accessible enough to be taught in high school classrooms across the country.

This inclusion is needed to reach broad audiences. It's logical that our most popular books are not complex, and I would not expect the future of popular reading to revert back to the lengthy sentences of George Washington's first State of the Union Address. Kerouac's most popular book, *On the Road*, scores at a reading level of 6.6 on the Flesch-Kincaid scale. And while I don't think Kerouac was referring to sentence structure when he said it, I still think that the following line is worth considering in this discussion: "One day I will find the right words, and they will be simple."

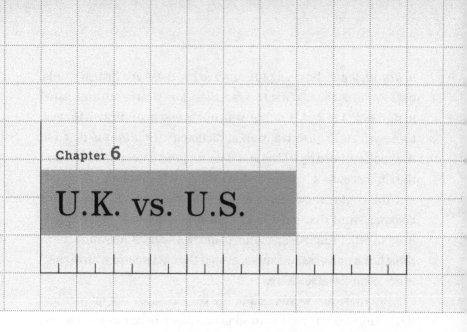

Chapter 6

U.K. vs. U.S.

*England and America are two countries
divided by a common language.*
—GEORGE BERNARD SHAW

A Tale of Wizards, Blokes, and Knickers

For many young American readers Harry Potter introduced them to a new world. I'm not just referring to the magic that takes place at Hogwarts, but the wonderful world of British English. For every fictitious word like *Muggle* for American children to learn, there was a British one like *bloke*. For every magic spell like *wingardium leviosa*, there was a British curse like *blimey*. Left and right, characters were *snogging* by common-room fireplaces. It was not just the magic that drew American readers in, but the *brilliant* British language.

And when we look at the data, we can actually *see* this: British words became so entangled in American readers' memories of Harry Potter that it changed the way they saw the characters, compared to British readers.

Let's consider those three British B-words: *bloke*, *blimey*, and *brilliant*. It may be a simplification to say these are British words, but I would not be the first to do so. *Bloke* and *blimey* are both listed in the book *A to Zed, A to Zee*, which is in essence a British English to American English translation dictionary. *Brilliant* is listed in the book *Divided by a Common Language* as a word "Not to Say in the US" because it "can mean something quite different in the US."

You could take exception to any of these words: *Blimey* is a cockney word that most in Britain would not use. *Bloke* is also used in other English-speaking countries such as Australia. *Brilliant* has a universal meaning, even if the exclamatory "Brilliant!" is not common elsewhere.

But these are words that at the very least are not quintessentially American. If you heard someone describe another person as a *"Poor bloke. Brilliant mind,"* you'd guess those words were spoken by someone from the U.K., not the U.S. And those exact words happen to be spoken by Hagrid in the first Harry Potter book to describe Professor Quirrell.

Are these words actually favored by British writers if we look at the numbers? Do all authors from the same side of the Atlantic have a shared style? Without looking at the data, it's hard to tell how many of the phrases and words we associate with either American or British writing are real and which are exaggerated stereotypes.

On a statistical level, the stereotype holds up when we turn to the data to look into these B-words. If you look at the British National Corpus (1980–1993) and the Corpus of Contemporary American English (1990–2015) you can see the disparity. These are both curated collections of hundreds of millions of words, meant to serve as benchmarks of the language as it's put into practice on either side of the Atlantic. In these samples *bloke* is used 27 times more often in British writing than American, *blimey* 30 times more often, and *brilliant* 45 times more often.

But I wanted to go a step further, looking at how these differences work their way into readers' minds. Let's start with *bloke*. Though used more in Britain, it's still not a big part of common language. In the British National Corpus *bloke* showed up just 1.2 times for every 100,000 words. However, Americans used it 0.045 per 100,000, perhaps making it attention-grabbing on the rare occasion it is read or heard. In the seven Harry Potter books, J. K. Rowling uses this word even more often than the average Brit, at a rate of just under 3 per 100,000 words—and it stuck out to American readers.

To find out how notable it was, I decided to compare British and American usage of *bloke* when writers are trying their best to imitate Rowling. I downloaded all Harry Potter stories set in the "Hogwarts Universe" that were of novel length (60,000-plus words) from FanFiction.net. Of these, 144 were by writers who listed their location as Great Britain and 555 were by writers who identified as American.

These are not ordinary writers or ordinary Harry Potter fans. These are people who have written at least 60,000 words—just 20% shorter than the first Harry Potter book—with the same characters and backdrop of Rowling's books. These are the diehards among diehards.

Your first instinct might be to think that, if these words are never used by Americans in ordinary speech, then they would not be used in American Harry Potter tributes. But the *opposite* is true. More American fans of Harry Potter were overindulgent in their use of *bloke* than British writers.

While just 10% of the Harry Potter fan fiction by people from Great Britain used *bloke* more than 3 times per 100,000 words, almost one-fourth of all American written fan fiction did.

One fan, an American, used it at a rate of over 60 times per 100,000 words. That's twenty times as often as Rowling used it. Despite the fact that *bloke* is even less common stateside, more Americans were obsessed with using the word than Brits.

The same is true for *blimey*.

Again, the worst offenders were stateside. One American (different than the *bloke* fan above) used *blimey* at a rate of more than 60 times per 100,000 words. American fan fiction used *blimey* at a higher rate than Rowling and twice as often as British fan fiction.

Though not to the same degree, even Americans used *bril-*

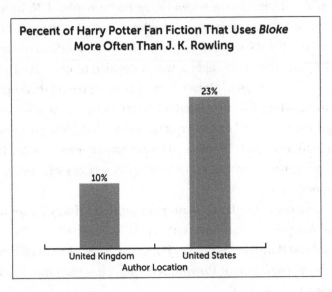

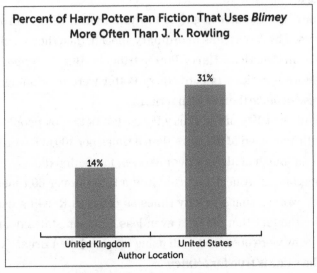

liant more. (If you were to look at just uses of *Brilliant!*, to exclude all uses of brilliant that are not exclamatory, the differential between U.S. and U.K. holds to within 1%.)

Once again the writer who used *brilliant* the most was an American.

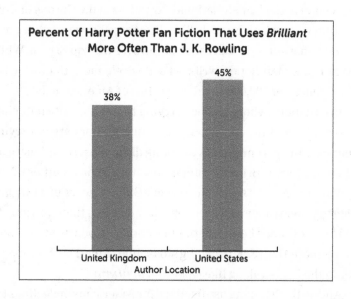

If there is one direct takeaway from this example it's that American fans who love Harry Potter love acting British—the Britishness of Hogwarts is part of Rowling's magical world for American readers. Because of the large discrepancies and the size of the sample (89 million words across 699 full-length fan-fiction books), we know these patterns are not random. The easiest way to sound British, or at the least *stereotypically* British, was to throw in some *blimey*s or *brilliant*s or *bloke*s.

But this is also a reminder that when comparing a loosely defined group of writers against each other, such as "British writers" versus "American writers," the group of texts examined is important. Someone who did not have context when looking at these two samples of text might assume that *blimey* is an American word.

In the example of fan fiction, it's interesting to note that less stereotypical signatures of British English don't make it into American writers' fan fiction. For example, British writers of Harry Potter fan fiction use *surely* more than three times as much as Americans writers. That's not surprising, as *surely* is almost twice as common in the British National Corpus as in the Corpus of Contemporary American English. The difference between *surely* and *blimey* is that *surely* doesn't stick out as over-the-top British. While 40% of British fan-fiction writers used *surely* more than Rowling did (10 times per 100,000 words), just 18% of Americans did.

In situations where no one is trying to imitate British English the results are different. When fan-fiction authors are not trying to emulate British characters the slang disappears from American writers *and* from British writers. Suzanne Collins, author of the popular young-adult series the Hunger Games, never used *blimey* or *bloke* in her trilogy and used *brilliant* 40% less than Rowling. In 420 Hunger Games book-length fan fictions no one used the word *blimey* more than J. K. Rowling's baseline. Americans no longer overdo the British slang like *bloke* and *brilliant*.

And as British slang use declines for Americans, it declines for those who identify as being from Great Britain as well. While 38% of British Harry Potter fan fiction used *brilliant* more than Rowling, 13% of British Hunger Games fan fiction does. And while 10% of British Harry Potter fan fiction uses *bloke* more than the Harry Potter author herself, less than 1.5% of British Hunger Games fan fiction does. Just as American fan-fiction writers delighted in

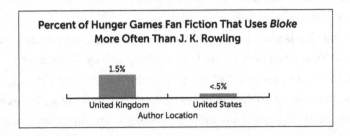

Percent of Hunger Games Fan Fiction That Uses *Bloke* More Often Than J. K. Rowling

1.5%

<.5%

United Kingdom   United States
Author Location

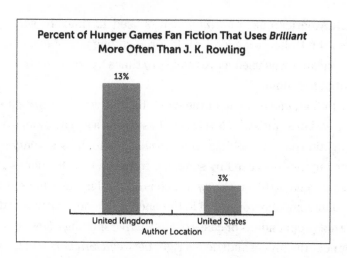

Percent of Hunger Games Fan Fiction That Uses *Brilliant* More Often Than J. K. Rowling

13%

3%

United Kingdom   United States
Author Location

adopting their favorite Britishisms, writers from the U.K. (for the most part) knew when to downplay them. No matter where you're from, no one imagines Katniss Everdeen having inner monologues where she worries about saving the blokes from District 12.

We can learn a lot from looking at what writers play up when trying to imitate authors from across the pond. But what differences do we find between American and British vocabularies when writers aren't focusing on imitating? What happens when they have other things on their mind—like, say, *sex*?

It turns out that erotica offers a perfect sample of text for comparing the two nations. I downloaded every single sex story from Literotica.com to see what differences we'd find. Based on the author's stated location, I placed approximately 3,200 stories as being by authors from the United Kingdom and about 15,000 as by Americans. This amounts to 76 million words of pure filth.

For every two times an American used the word *brilliant* in a sex story, someone from the United Kingdom used it three times. The word *surely* was used twice as much by writers from the United Kingdom compared to those from America. And for each American use of *blimey* it was used 12 times by someone from the United Kingdom, while for each one use of *bloke* by an American it was used more than *fifty* times by someone from the United Kingdom.

In fact, *bloke* is one of the most distinct words of British sex stories. On a statistical level it's not as strong a giveaway as words with alternative spellings like *humour*, which is seventy-plus times more often used by someone from the U.K. It's not as distinct as a name like *Charley*, which is sixty-plus times more often used by someone from the U.K. But once you remove all words that are not proper nouns or cases of alternative spellings, *bloke* comes out near the top—sandwiched right between *knickers* and *lads*.

Here are the others (feel free to let your imagination go wild):

Most Distinctive Words: British vs. American Erotica	
UNITED STATES	UNITED KINGDOM
Comforter	Wanked
Trailer	Knickers
Nightstand	Bloke
Restroom	Lads
Cowboy	Suspender
Semester	Settee
Grade	Shagging
Dr.	Fancied
Motel	Sod
Cops	Bum
Closet	Loo
Railing	Toilets
Downtown	Hugely
Tub	Squashed
Parking	Cope
Scooted	Duvet
Ranch	Sordid
Refrigerator	Pub
Truck	Corridor

Of course, these lists are by definition relative. The numbers do not imply that all American erotica authors are obsessed with trailers. Words in the list above do not have to appear in abundance to be a giveaway of an author's origin; they just have to appear much more often relative to the texts they are being compared against.

If you break down the results by smaller regions you get different results. While writers from Texas (this is stated author location, not location of the story) use the word *trailer*, those from New York are a lot more interested in what happens in the *subway*.

Most Distinctive Words: New York vs. Texan Erotica	
TEXAS	NEW YORK
Ranked	Subway
Trailer	Popsicle
Soldiers	Senator
Sergeant	Butthole
Bunk	Museum
Arena	Landlord
Evidently	Sin
Altar	Jacuzzi
Alley	Thrusted
Captain	Shrugs

There are also great discrepancies in the defaults each author uses as their point for comparison. In an American's world, a typical person around them doesn't have an "American accent." They just talk like everyone else, so there's no need to describe it. I looked at all instances of *accent* in the collection of tens of thousands of stories to see what type of accents were being described. Not all characters in these stories are paramours, but I think it's fair to assume a great deal of the accents mentioned were describing one love interest or another. Some accents find their way into more sex stories, out of being a combination of noteworthy and appealing. The most common accents in Literotica follow.

Most Popular Accents in Erotica

1. Southern	6. Irish
2. French	7. Spanish
3. British	8. Australian
4. English	9. European
5. American	

Below, I've broken the list down by author location. As you would expect, American authors seldom describe their own characters as having an American accent. Likewise for British authors. Yet "American accent" tops the British list, and "British accent" clocks in at second on the American list (right after the sultry Southern drawl).

Most Popular Accents in Erotica by Author Location	
AMERICAN AUTHOR	BRITISH AUTHOR
Southern	American
British	English
English	French
French	Irish
European	European
American	Scottish
Italian	Australian
Spanish	Southern
Irish	London

Read Local

How important is home-field advantage for writers? Do American or British authors perform better away from their base?

The other sections in this chapter have focused on writing style, but I was also curious about how authors from different markets are received. If you are trying to write a global bestseller, is it easier for an American to sell in Britain or the other way around? Before we go back to the difference in popularity of American and British authors in different markets, consider the smaller example of Stephen King.

I've written about King many times in this book for different reasons. He has authored dozens of bestsellers over the span of decades, which makes him a good target of analysis, and he has also written about writing. Among book lovers, he's one of the most popular living authors. On Goodreads.com, a social networking site designed to be used by avid readers, he has more "followers" than any other author.

But I will be the first to admit that I also write about Stephen King so often because of my personal bias. Stephen King used to live in Bridgton, Maine, and has used the small town as the backdrop of several stories. *The Mist* is about the rise of mysterious monsters from a storm over Long Lake in Bridgton. As I write this chapter I am sitting in a house overlooking Long Lake. I picked *The Mist* out years ago to read for this very reason. For me, there was an added element of fun to read a book set in a location I knew, written by a local author (no matter how global his success is).

King is still a local legend around Bridgton. Thirty-four of King's fifty-one* novels take place in Maine. And reading a Stephen King novel in New England makes the book have a local feel even though he's a mega-author. It's like rooting for the Boston Red Sox—a massive business that still feels like the local hero.

And the data backs this up. Though he is the most popular author on Goodreads, King loses out to many others on the more populist metric of Facebook fans. As of writing this, King has 4.5 million American fans, but the romance novelist Nicholas Sparks

---

* This is based on my count. Books have multiple settings. This includes all books that have any action taking place in Maine.

has around 25% more fans, at 5.8 million. The numbers vary, however, by region and state. In Mississippi and Alabama, Sparks has 75% more fans than King. In his home state of North Carolina, the *The Notebook* author has 60% more fans.

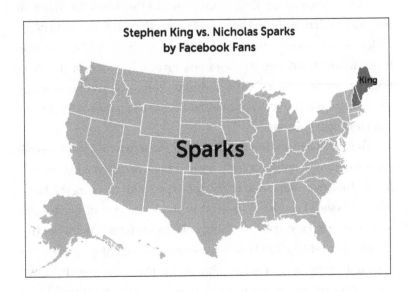

Stephen King vs. Nicholas Sparks
by Facebook Fans

But in King's home territory of Maine, King is 20% more popular (measuring by Facebook likes) than Sparks. In neighboring New Hampshire, King edges out Sparks as well.

The differences in state-to-state fandom are minimal compared to the differences from nation to nation. In 2014 Amazon decided to publish a list of "100 Books to Read in a Lifetime." They published one version on their American Amazon.com and a different version on Amazon.co.uk. There were just 21 books that both the American and British list agreed on.

The biggest discrepancy between the lists is easy to spot even just scrolling through. The American list had 69 books written by American authors, while 16 were by authors from the United Kingdom. The British list had 70 books written by people from the

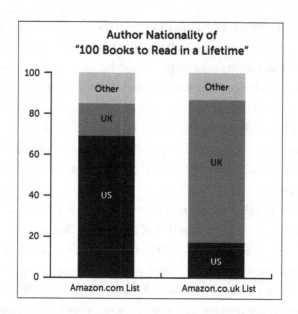

Author Nationality of "100 Books to Read in a Lifetime"

United Kingdom and 17 by Americans. By a near identical ratio, the vast majority of books were by authors from the same country as the list maker.

A list published by the London-based *Telegraph* in 2015 titled "100 Novels Everyone Should Read" featured 43 books by authors from the United Kingdom and just 16 by Americans. American critics Richard Lacayo and Lev Grossman put together a list for the American magazine *Time* called "All-Time 100 Novels." It featured 59 Americans and just 34 Brits.*

Pop culture has been called by many America's greatest export, and in Britain 84%† of the movie market is dominated by American films. It's obvious that American authors don't have the same overwhelming market share in Britain as American filmmakers, but

---

* For these numbers and the chart above, author nationality is not always singular or certain. The country the author identified with or spent most time writing in was selected.

† According to 2014 article "The American Cinema: A Cultural Imperialism?"

are Americans eroding Britain's preference for its own writers? I decided to take a look at the bestseller lists from each country.

Let's start with Stephen King, who has achieved international success. He's had 34 number one *New York Times* bestsellers. In the same time period he's seen 19 (as of 2014) number one best-sellers on *The Sunday Times* list in Britain. While in the States he's beaten by only James Patterson for the number of number one hits, in the U.K. he's bested by British authors Catherine Cookson, Terry Pratchett, and Dick Francis.

While King's success is tempered in Britain, he's still a huge force. There are only a handful of successful British writers who remain more popular than him in Britain. On the other hand, the same British writers who have more hits than him (Cookson, Pratchett, and Francis) in the U.K. have found much less success stateside. None of those three authors has ever had a number one *New York Times* hit.

Maybe you don't think books should be measured by a pop-ularity contest alone, but if they were the Americans are gaining fast. *The Sunday Times* started publishing their bestseller list in 1974. Each week listed the top ten entries in many categories in-cluding Hardback—Fiction. In *The Sunday Times*' first year of bestsellers Brits beat out the Americans without a sweat. Looking at just those who were from either nation, 84% were British to 16% on the American side in 1974.

Ten years later Americans had inched up to 22%. In 1984 they had inched up to 27% and another decade later to 33%. In 2014, forty years after the list's first year, Americans were at a ratio of 37% to British writers' 63%.

The fraction of American books reaching the bestseller list has more than doubled in the last forty years. They are still outnum-bered, but at 37% they represent a huge portion of the market—an American invasion.

In the same time, the rate of British bestsellers in America has

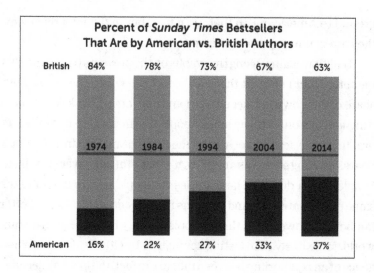

Percent of *Sunday Times* Bestsellers That Are by American vs. British Authors

British	84%	78%	73%	67%	63%
	1974	1984	1994	2004	2014
American	16%	22%	27%	33%	37%

been falling. In 1974 books by British authors were about 38% of the *New York Times* bestseller list. That's about what American authors are in Britain now. In 2014 British authors made up only 11% of the list—less than American writers did in Britain in 1974.

Should Brits worry that they are using losing their grasp on the language they created? Not only are British writers losing huge

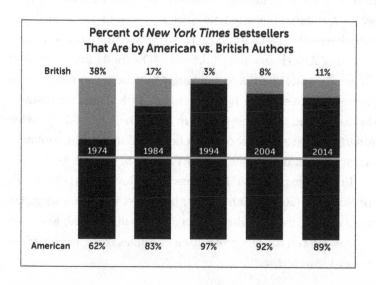

Percent of *New York Times* Bestsellers That Are by American vs. British Authors

British	38%	17%	3%	8%	11%
	1974	1984	1994	2004	2014
American	62%	83%	97%	92%	89%

ground to Americans on their own turf, but in the long-term view they are coming up short overseas.

I caution against using the graphs on the previous page to predict the future, but I suspect that the days of Brits claiming such a huge share of their own market are over and not coming back. The United Kingdom has one-fifth as many people as America, so it is only natural that America did not stay as the nonfactor it was in 1974. But if Sparks vs. King shows us anything, it's that a full takeover is unlikely. People have a desire to hear stories by people who understand the same places, interests, and conflicts that they do. And if Harry Potter fan fiction shows us anything, people take joy in exploring a distant world. More likely than lasting dominance by either side is the emergence of an equilibrium between the two. In fact, the last few decades seem to bear that out, as the percentage of British fiction in America has bounced back from a low of 3% in 1994 to 8% in 2004, and then to 11% in 2014 (thanks in part to a new world of wizardly blokes).

Are Americans Loud?

British people use the word brilliant *often*. That's a stereotype—which the data shows is true. Just as we've focused on certain Britishisms in this chapter, there are words and phrases stereotypical of Americans, like "take-out," "I'm good," and "heads-up." These are small quirks in language, but there are also grander stereotypes describing how Americans talk. Take, for instance, the stereotype that Americans are loud. The loud and obnoxious American tourist has become a cliché in Europe. But, looking at literature, could it be true that Americans, in general, are loud?

In 2014 a paper titled "Loudness in the Novel" was published by the Stanford Literary Lab. The author, Holst Katsma, categorized "speaking verbs" into three categories: loud, neutral, and quiet. Katsma offered the following three examples from *Alice's Adventures in Wonderland*:

Loud: "Off with their heads!" **shouted** the Queen.

Neutral: "I suppose so," **said** Alice.

Quiet: He **whispered**, "She's under sentence of execution."

Katsma was able to explore a few trends with the data, including finding which words were associated with loud dialogue more than neutral dialogue. He used old novels (ones from the 1800s) and found in his sample that *why*, *stop*, *God*, and the exclamation point were found in loud dialogue at a higher rate than neutral dialogue. Words like *night*, *well*, *suppose*, and the period were found in neutral dialogue more.

I wanted to find out if Katsma's method could help answer the question of whether Americans *write* loudly. Can we measure whether American English is itself louder than British English?

Below are the words that Katsma classified as "loud," "neutral," or "quiet." I used his list for my own analysis, though I did add first-person present and third-person present verbs to the mix (to account for different narrators and tenses).

LOUD	NEUTRAL	QUIET
Cried, Exclaimed, Shouted, Roared, Screamed, Shrieked, Vociferated, Bawled, Called, Ejaculated, Retorted, Proclaimed, Announced, Protested, Accosted, Declared	Said, Replied, Observed, Rejoined, Asked, Answered, Returned, Repeated, Remarked, Enquired, Responded, Suggested, Explained, Uttered, Mentioned	Whispered, Murmured, Sighed, Grumbled, Mumbled, Muttered, Whimpered, Hushed, Faltered, Stammered, Trembled, Gasped, Shuddered

Any "loudness score" will be imperfect. It depends on defining words like *shouted* and *proclaimed* as (equally) loud and *sighed* or *grumbled* as (equally) quiet. Not every author describes every piece of dialogue with a colorful descriptor. And Katsma's extensive list does not include all verbs ever used to describe "loud" or "quiet" dialogue. But his method offers a workable metric, especially in aggregate, so I've done some tests using the exact words listed above.

Authors do not write every book at the same level, but their habits tend to be consistent. Below is a plot showing the difference in "loud" and "quiet" speaking verbs in J. K. Rowling's Harry Potter series and Dan Brown's Robert Langdon series. The third Harry Potter book was 51% "loud" to 49% "quiet," meaning Rowling used about as many "loud" verbs as "quiet" ones. *Angels and Demons* comes in at 73% "loud" to 27% "quiet," meaning that in this book Brown used almost three times as many "loud" verbs as "quiet" ones.

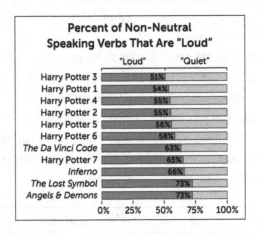

On the opposite page is a chart showing the loudness of the fifty different authors that I've used as examples throughout this book. Black bars denote British authors, while gray bars represent Americans.

There are a fair amount of British authors on the quiet end—but it's far from consistent. If anything, the more striking finding is how far E L James and Stephenie Meyer diverge from the rest of the authors here—in large part because of personal style and the focus on romance in their books. There are some additional surprises: the Harry Potter series and Agatha Christie novels are so filled with action that I would not have guessed them to be on the "quiet" end of the spectrum. Or that Hemingway is the loudest of the loud (though maybe that's not *too* surprising).

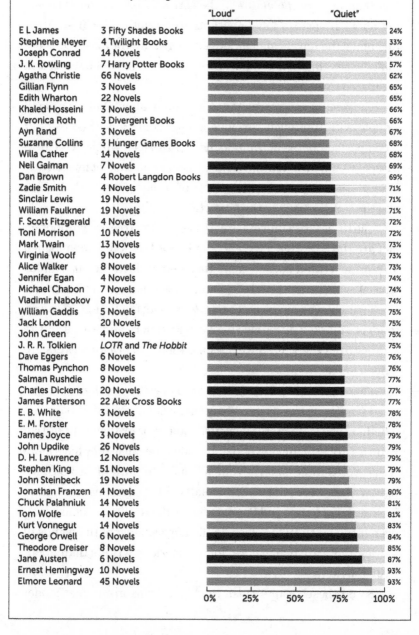

Percent of Non-Neutral
Speaking Verbs That Are "Loud"

		"Loud"	"Quiet"	
E L James	3 Fifty Shades Books			24%
Stephenie Meyer	4 Twilight Books			33%
Joseph Conrad	14 Novels			54%
J. K. Rowling	7 Harry Potter Books			57%
Agatha Christie	66 Novels			62%
Gillian Flynn	3 Novels			65%
Edith Wharton	22 Novels			65%
Khaled Hosseini	3 Novels			66%
Veronica Roth	3 Divergent Books			66%
Ayn Rand	3 Novels			67%
Suzanne Collins	3 Hunger Games Books			68%
Willa Cather	14 Novels			68%
Neil Gaiman	7 Novels			69%
Dan Brown	4 Robert Langdon Books			69%
Zadie Smith	4 Novels			71%
Sinclair Lewis	19 Novels			71%
William Faulkner	19 Novels			71%
F. Scott Fitzgerald	4 Novels			72%
Toni Morrison	10 Novels			72%
Mark Twain	13 Novels			73%
Virginia Woolf	9 Novels			73%
Alice Walker	8 Novels			73%
Jennifer Egan	4 Novels			74%
Michael Chabon	7 Novels			74%
Vladimir Nabokov	8 Novels			74%
William Gaddis	5 Novels			75%
Jack London	20 Novels			75%
John Green	4 Novels			75%
J. R. R. Tolkien	*LOTR* and *The Hobbit*			75%
Dave Eggers	6 Novels			76%
Thomas Pynchon	8 Novels			76%
Salman Rushdie	9 Novels			77%
Charles Dickens	20 Novels			77%
James Patterson	22 Alex Cross Books			77%
E. B. White	3 Novels			78%
E. M. Forster	6 Novels			78%
James Joyce	3 Novels			79%
John Updike	26 Novels			79%
D. H. Lawrence	12 Novels			79%
Stephen King	51 Novels			79%
John Steinbeck	19 Novels			79%
Jonathan Franzen	4 Novels			80%
Chuck Palahniuk	14 Novels			81%
Tom Wolfe	4 Novels			81%
Kurt Vonnegut	14 Novels			83%
George Orwell	6 Novels			84%
Theodore Dreiser	8 Novels			85%
Jane Austen	6 Novels			87%
Ernest Hemingway	10 Novels			93%
Elmore Leonard	45 Novels			93%

0%   25%   50%   75%   100%

If you are familiar with the Rowling novels, you might start to hypothesize about what contributes to their "quietness." Does the fact that much of the action in the series revolves around sneaking about Hogwarts have anything to do with this? If so, does this make the novels appear more "quiet" than they are? Or is the "quiet" sneaking action a perfect example of the subdued action compared to an American series?

The results are interesting to speculate about, but it's a small sample with too much variance among genre and time to offer a comparison between Americans or Brits as a whole. The ideal sample would be narrow in terms of genre and subject matter, and in terms of time period—with plenty of authors from both the U.K. and the U.S. So, back to the fan fiction.

The three story universes with the most entries on FanFiction .net are Harry Potter, Twilight, and Percy Jackson. I downloaded all stories written in each genre that were full-length (60,000-plus words), which amounted to 284 million words over 2,225 stories.

Twilight was on the quiet side. Within its fan fiction, however, there is a marked difference between its American writers and its writers from the United Kingdom. Overall American writers are indeed louder.

There are many ways we could slice the results. I categorized fan-fiction novels into being either "loud" or "quiet" based on whether they were more than 50% "loud." For example, all of Dan Brown's Robert Langdon books are "loud" because they have more "loud" verbs than "quiet." While around one-third of British Twilight writers were loud, over one-half of all Americans were. If we look at Harry Potter and Percy Jackson, the results are consistent, even if the magnitude is less. In the huge sample of thousands of stories, there are many more "loud" American writers than "loud" British writers.

So do these results mean that Americans are in fact louder

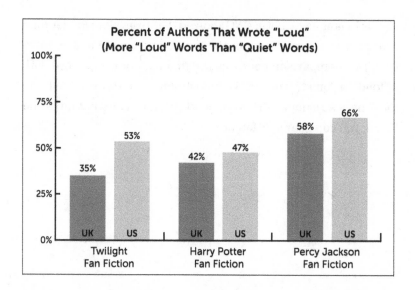

Percent of Authors That Wrote "Loud"
(More "Loud" Words Than "Quiet" Words)

in their writing than the quiet Brits? I would argue that it does. In samples of hundreds of millions of words by thousands of individuals Americans come out "louder" than Brits.

The fifty writers I've been looking at showed no conclusive results, but if you look at other large fiction samples, the small but noticeable pattern we find in fan-fiction stays true. I took the British National Corpus (BNC) and the Corpus of Contemporary American English (COCA) and looked at how "loud" and "quiet" words were used. These samples don't allow for as perfect a comparison as fan fiction, but they are similar. Both corpuses contain hundreds of millions of words and both cover a similar modern time period (1980–1993 for BNC and 1990–2015 for COCA). I looked at just the "fiction" texts in each sample. Because the corpuses are based on texts that the researchers from different institutions choose to include, it's possible one could end up having more thrillers, or more romances, or more young adult, or more literary fiction.

Still they are huge samples that are used to compare general trends in language. If the BNC showed that Brits were louder than Americans, you might doubt the fan-fiction sample above. However, this comparison confirms the finding. For every word with a "loud" or "quiet" value in the British sample, 66% were "loud." In the American sample, 73% were "loud." It's not deafening, but if you listen, you can hear a difference.

Chapter 7

Clichés, Repeats, and Favorites

"Don't you wish sometimes that writing were just like sports? That you could just go out there and see who'd win? See who's better. Measurably. With all the stats."
—DUB TRAYNOR IN *THE INFORMATION* BY MARTIN AMIS

When I was a ten-year-old I wrote a series of eight superhero "books," each taking up between 40 to 100 handwritten pages. There were two protagonists, one of whom was named *Bubonic Boben Blaster* after myself (*Bubonic Bo**ben** Blaster*). He was like me in every way except for his one superpower: the ability to infect his enemies with the bubonic plague.

It was bad.

Here's a passage (bolded for emphasis):

Then Bubonic Boben Blaster took over control of the airplane. **Then** he kept flying until he saw grass and he put the plane on autoland. **Then** Bubonic Boben Blaster helped everyone jump off the plane with their luggage. No one knew where they were. **Then** Bubonic Boben Blaster looked at a map and saw they were in South Carolina.

I remember the day in the year 2000 when I read this page to my fourth-grade class. Of all the possible feedback a teacher could have offered up after that passage, I was told: "Do not start two sentences in a row with the same word."

For years this advice stuck with me, and it became a simple rule that I followed in essays. Changing the first word around was a surefire way to make sure consecutive sentences did not have identical structure. Many writing guides offer the same advice.

But if you're not a ten-year-old with weak control of the written word, it's clear that repeated words can have strong rhetorical power. While the recurring *Then* in my own childhood writing is cringeworthy, in other cases repetition has a clear rhetorical purpose. Consider the famous line below, from one of Winston Churchill's World War II speeches.

> **We shall** go on to the end. . . . **We shall** fight on the beaches, **we shall** fight on the landing grounds, **we shall** fight in the fields and in the streets, **we shall** fight in the hills; **we shall** never surrender. . . .

The repetition is what makes it memorable. When done well such repetition allows for a sense of rhythm and power. Here's a literary example from Charles Dickens's *Hard Times* with the repeated sentence beginnings in bold.

> He was a rich man: banker, merchant, manufacturer, and what not. A big, loud man, with a stare, and a metallic laugh. **A man** made out of a coarse material, which seemed to have been stretched to make so much of him. **A man** with a great puffed head and forehead, swelled veins in his temples, and such a strained skin to his face that it seemed to hold his eyes open, and lift his eyebrows up. **A man** with a pervading appearance on him of being inflated like a balloon, and

ready to start. **A man** who could never sufficiently vaunt himself a self-made man. **A man** who was always proclaiming, through that brassy speaking-trumpet of a voice of his, his old ignorance and his old poverty. **A man** who was the Bully of humility.

The technical term for this device—repeating a word or phrase at the beginning of consecutive sentences—is *anaphora*. And Dickens is the master. You may be familiar with the opening to *A Tale of Two Cities*—"It was the best of times, it was the worst of times . . ."—which is one of the most memorable examples of anaphora in all of English literature. His longest string of sentences starting with the same word lasts 26 sentences in a row, all starting with *when* in his novel *A Haunted Man*.

So far, this book has been filled with big-picture perspectives on whole genres and collections of authors, deep dives into overlooked but universal elements such as word frequencies and sentence lengths. But sometimes the unique quirks, whether they be special words or literary devices, are what make a reader remember an author's voice. Data can shed light on these smaller questions as well. There are rules for writing, but every good rule has been broken by a good author. Anaphora is one of those rules.

In this chapter I'll be looking at writers' eccentricities. In honor of Dickens and the opening of *A Tale of Two Cities*, I'll start by presenting a data-driven case study on anaphora: a tale of two columnists.

Krugman vs. Brooks

Paul Krugman is a left-leaning professor and Nobel Prize winner. David Brooks is a right-leaning journalist and author of several books. Both are noted for their opinion columns in the *New York Times*.

I collected the most recent year's worth of opinion columns from both Krugman and Brooks. Next time you read one of Brooks's columns, keep your eyes open for anaphora. In 91 out of all 93 David Brooks columns from that year-long period, he had at least two sentences in a row that started with the same word. It's rare for Brooks not to employ anaphora in some form. Here's the start of one of his columns, a short but pointed example: *Some people like to keep a journal. Some people think it's a bad idea.*

Or consider these six consecutive sentences that all start off with the same four words:

> **It used to be** that senators didn't go out campaigning against one another. **It used to be** they didn't filibuster except in rare circumstances. **It used to be** they didn't block presidential nominations routinely.
>
> **It used to be** that presidents didn't push the limits of executive authority by redefining the residency status of millions of people without congressional approval. **It used to be** that presidents didn't go out negotiating arms control treaties in a way that doesn't require Senate ratification. **It used to be** that senators didn't write letters to hostile nations while their own president was negotiating with them.

In total 9% of all Brooks sentences start with the same word as the previous sentence. This is a rough metric for anaphora, but an informative one. In contrast to Brooks, only 2% of Krugman's sentences fall into that criterion.

While over 95% of all Brooks columns had at least one example of anaphora, less than half of Krugman opinion columns did. It's not like Krugman and his editor avoid anaphora like the (bubonic) plague, but it's rare. When he does start off consecutive sentences with the same word it's rarely the rhetorical choice that it is for Brooks.

If I were to offer my own reaching theory as to why Krugman shies away from the strong rhetoric it would be from his academic background. In a need to be comprehensive, he hedges, negotiates, and qualifies all his points. The most common words that Krugman and Brooks use to start their sentences offer evidence of this theory.

David Brooks uses *the* as his most frequent sentence starter. This is to be expected. Though sometimes bested by *he*, *she*, or *I* in literature, *the* is almost always the most common sentence lead across writing. Pronouns aside, it's hard to think of any other word that could top *the*. Brooks uses *the* more than twice as often to start sentences as any other individual word.

But what about Krugman? He starts more sentences with *but* than *the*. The conjunction *but*, a word that indicates the writer is about to say something to undermine or qualify a previous statement in some way, is Krugman's favorite way to start a sentence. While Brooks uses *the* twice as often as *but* to start a sentence, Krugman uses *but* a full 33% more often than *the*.

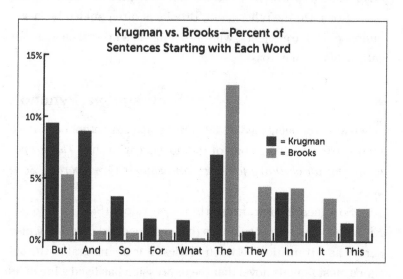

And in the most common three-word sentence openings found in Krugman and Brooks columns, we see Krugman clarifying and chaining his sentences while Brooks favors a more direct approach.

Top Three-Word Sentence Openers: Krugman vs. Brooks	
KRUGMAN	**BROOKS**
It's true that	Over the past
At this point	Most of us
Which brings me	If you are
The point is	You have to
As a result	It is a
The truth is	This is the
As I said	In X% of
On the contrary	On the other
One answer is	The people who
The answer is	In the first

Krugman's affinity for *But* or *So* might be part of the reason he cannot use anaphora. There are not many ways to begin consecutive sentences with *But* without devolving into a tangle of contradictions. On the other hand, Brooks is often guided by larger philosophical arguments, where rhetoric is a crucial element for getting his point across.

Vonnegut vs. Pynchon

When Kurt Vonnegut died in 2007 *Time* magazine author Lev Grossman started off his eulogy of the novelist with this: *The proper length for an obituary for Kurt Vonnegut is three words: "So it goes."*

"So it goes" is a refrain used by the narrator in *Slaughterhouse-Five*. Vonnegut uses it to tell us that someone is dying, and it also serves as a transition between stories. It is so integral to the author's most famous novel that the expression has lived a life of its own, as seen in Vonnegut's obituaries and the title of Vonnegut's biography by Charles J. Shields.

From a literary point of view the repetition of "So it goes" helps set the tone of the entire story. But the phrase also has a unique property from a mathematical point of view: With 106 uses, it is the most frequent sentence in all of Vonnegut's novels.

And in my survey of all fifty authors, no other sentence is used as often in a single work. In fact, no other sentence comes close. By my count the second-place sentence comes in at just 35 uses. That sentence? "And so on." This is used by Vonnegut again in his novel *Breakfast of Champions* to similar effect.

At more than 100 uses, "So it goes" represents a measurable portion of the entirety of *Slaughterhouse-Five*. It accounts for 2.5% of the sentences in the entire book—about one out of every forty.

The repetition of "So it goes" in *Slaughterhouse-Five* is different from the anaphora discussed above. It's not a phrase that's used to open consecutive sentences. However, it is emblematic of Vonnegut's approach to writing. He used repetition and anaphora often. More than 12% of all sentences in Vonnegut's *Slaughterhouse-Five* start with the same word as the previous sentence. Among all the classic and popular novels surveyed, this ranks toward the top.

On the following page the top ten books are ranked when looking at this very basic approximation of anaphora—the percentage of sentences that begin with the same word as the previous sentence.

Most of these names should not surprise you. Virginia Woolf's *The Waves* is an experimental novel written in soliloquies. In his book *Survivor*, Chuck Palahniuk writes: "There are only patterns, patterns on top of patterns, patterns that affect other patterns. Patterns hidden by patterns. Patterns within patterns." It's no surprise to see some of his work near the top of this list.

Books with the Most One-Word Anaphora

BOOK	AUTHOR	ANAPHORA %
The Waves	Virginia Woolf	16.0
Survivor	Chuck Palahniuk	13.5
The Ocean at the End of the Lane	Neil Gaiman	13.3
Breakfast of Champions	Kurt Vonnegut	13.2
Slaughterhouse-Five	Kurt Vonnegut	12.3
The Fountainhead	Ayn Rand	12.3
Cat and Mouse	James Patterson	11.9
Slapstick	Kurt Vonnegut	11.5
Lullaby	Chuck Palahniuk	11.4
Kiss the Girls	James Patterson	11.3

"So it goes" aside, Vonnegut speaks in repetition often, and to great effect. Here's a passage from *Cat's Cradle*:

> Before we took the measure of each other's passions, however, **we talked** about Frank Hoenikker, and **we talked** about the old man, and **we talked** a little about Asa Breed, and **we talked** about the General Forge and Foundry Company, and **we talked** about the Pope and birth control, about Hitler and the Jews. **We talked** about phonies. **We talked** about truth. **We talked** about gangsters; **we talked** about business. **We talked** about the nice poor people who went to the electric chair; and **we talked** about the rich bastards who didn't. **We talked** about religious people who had perversions. **We talked** about a lot of things.

The percentage of sentences in which the first two words are the same as the previous sentence—what you might call two-word anaphora—reveals a similar pattern to the chart above. It's not just that Vonnegut, Woolf, or Rand was being lazy and starting each sentence off with a boring *the*. Two-word anaphora tends not to end up

Books with the Most Two-Word Anaphora		
BOOK	AUTHOR	TWO-WORD ANAPHORA %
The Waves	Virginia Woolf	5.5
Survivor	Chuck Palahniuk	4.0
The Chimes	Charles Dickens	3.6
Fight Club	Chuck Palahniuk	3.3
Shalimar the Clown	Salman Rushdie	3.2
The Torrents of Spring	Ernest Hemingway	3.2
Slapstick	Kurt Vonnegut	3.2
The Sirens of Titan	Kurt Vonnegut	3.1
Atlas Shrugged	Ayn Rand	3.0
The Battle of Life	Charles Dickens	3.0

in the final text unless it is intentional. The same authors top the following list, with the arrival of the master himself, Dickens.

On the other end of the spectrum we can look at the type of book that chooses not to use anaphora very often. *Bleeding Edge* by Thomas Pynchon has over 10,000 sentences and just 1.6% of those (163) begin with the same word as the previous sentence. Pynchon's work makes up five of the top ten slots here.

Slaughterhouse-Five has 87 instances wherein three straight

Books with the Least One-Word Anaphora		
BOOK	AUTHOR	ANAPHORA %
Bleeding Edge	Thomas Pynchon	1.6
Vineland	Thomas Pynchon	1.9
Inherent Vice	Thomas Pynchon	2.2
Against the Day	Thomas Pynchon	2.3
The Children	Edith Wharton	2.4
A Son at the Front	Edith Wharton	2.5
Other Voices, Other Rooms	Truman Capote	2.6
Mason & Dixon	Thomas Pynchon	2.6
Twilight Sleep	Edith Wharton	2.7
The Grass Harp	Truman Capote	2.8

sentences start with the same word. If Pynchon were to write in the same exact manner, then *Bleeding Edge* would be expected (based on its length) to have 230-plus such cases of three consecutive sentences starting with the same word. Instead, there are just eight.

Pynchon favors variety; Vonnegut favors familiarity. I looked at the most common three-word strings to start off sentences in every book. In *Slaughterhouse-Five* the ten most common cases are:

1. So it goes
2. There was a
3. It was a
4. And so on
5. He was a
6. He had been
7. He had a
8. They had been
9. One of the
10. Now they were

These represent almost 7% of all sentences in the book. In *Inherent Vice* the ten most common openers are:

1. By the time
2. After a while
3. There was a
4. I don't know
5. It was a
6. Not to mention
7. What do you
8. I don't think
9. Doc must have
10. Now and then

These represent less than 1.5% of all sentences in the book.

So Pynchon's variety is seen not just in his lack of anaphora, but in the variety of his sentences as well. In the sample of all fifty authors seen in the previous chapter, Pynchon ranks near the bottom when we look at the percent of sentences that use his top ten openers.* Only James Joyce beats him for more variety. Vonnegut ranks toward the top, behind only a few authors—Hemingway,

---

* Each book's top ten sentence openers were calculated individually from the rest of their corpus.

Gaiman, Rand, Rowling, and Stephenie Meyer—who rely on their top ten openers more often.

For curiosity's sake here are the top ten most popular three-word sentence openers for a handful of other notable books. Like Pynchon and his openers, I was looking for variety so I hand-picked the examples to be as different as possible. Limited as they may be, each offers a tiny glimpse into the depths of each work.

FIGHT CLUB	FIFTY SHADES OF GREY	THE ADVENTURES OF TOM SAWYER
CHUCK PALAHNIUK	E L JAMES	MARK TWAIN
1. You wake up	1. Christian Grey CEO	1. By and by
2. This is the	2. I want to	2. There was a
3. I am Joe's	3. My inner goddess	3. I don't know
4. This is a	4. His voice is	4. It was a
5. I go to	5. From Christian Grey	5. What is it
6. You have to	6. I have to	6. There was no
7. This is how	7. I shake my	7. The old lady
8. The space monkey	8. I don't want	8. Then he said
9. The first rule	9. I need to	9. What did you
10. Tyler and I	10. I don't know	10. At last he

PRIDE AND PREJUDICE	THE GREAT GATSBY	THE OLD MAN AND THE SEA
JANE AUSTEN	F. SCOTT FITZGERALD	ERNEST HEMINGWAY
1. I do not	1. It was a	1. The old man
2. I can not	2. I want to	2. He did not
3. I am sure	3. There was a	3. I wish I
4. I am not	4. She looked at	4. I do not
5. It is a	5. I'm going to	5. But there was
6. Mrs. Bennet was	6. I'd like to	6. But I will
7. It was a	7. He looked at	7. But I have
8. I have not	8. She turned to	8. There was no
9. She is a	9. It was the	9. I wonder what
10. It was not	10. I don't think	10. The sun was

ANIMAL FARM	THE DA VINCI CODE	TO KILL A MOCKINGBIRD
GEORGE ORWELL	DAN BROWN	HARPER LEE
1. The animals were	1. The Holy Grail	1. Jem and I
2. It was a	2. I don't know	2. There was a
3. All the animals	3. It was a	3. I don't know
4. I do not	4. Find Robert Langdon	4. It was a
5. The animals had	5. Langdon felt a	5. There was no
6. None of the	6. There was a	6. Aunt Alexandra was
7. Do you not	7. O Draconian Devil	7. It was not
8. No animal shall	8. Oh lame saint	8. Judge Taylor was
9. Whatever goes upon	9. Langdon and Sophie	9. That's what I
10. As for the	10. The altar boy	10. I don't want

Too Many Clichés in the Kitchen

Martin Amis hates clichés. When the English novelist published a book of his literary criticism, he decided to call it *The War Against Cliché*. Amis explains his title by asserting that "all writing is a campaign against cliché. Not just clichés of the pen but clichés of the mind and clichés of the heart."

Amis is not alone. Clichés, by definition, are overused. No writer considers themselves an abuser of tired language. But what is considered a cliché is open to interpretation. Imagine if we had Amis read thousands of books at once to assign each an Amis Cliché Score. Modern technology is not advanced enough to create Martin Amises to do our work for us, but even if it were it would have limited value. It's very likely that Martin Amis, an academic from a literary family, would have different standards from anyone else.

And it's also hard to imagine weighing one cliché against another. A story that ends with the good guy saving the day and getting the girl has a clichéd ending. How do you weigh that clichéd ending with another book that has a more original ending but is full of stock characters? Which book is more "clichéd"? That's not a question an objective method could answer.

But if we focus on overused phrases, what Amis would consider "clichés of the pen," it might be possible to answer the narrow question of who writes with the most clichés. Which author uses sayings such as "fish out of water, "dressed to kill," or "burn the midnight oil" the most?

The above are all expressions found in *The Dictionary of Clichés*. The book by Christine Ammer published in 2013 contains over 4,000 clichés and, to my knowledge, is the largest collection of English language clichés. Because of its size, and the fact that Ammer had been compiling collections of clichés and colloquialisms for 25 years before *The Dictionary of Clichés* was published, I've chosen her as the official arbiter of what does or what does not count as a cliché.

Every author uses clichés, even if each has a unique voice. For the chart on the next page, I picked an assortment of authors and one cliché (from *The Dictionary of Clichés*) that each uses at an unusually high rate.

It's worth noting that Ammer's cliché criterion, even with 4,000 entries, is not comprehensive and may be different than Amis's list of expressions, or mine or yours. When detailing his view on clichés in an interview with Charlie Rose, Amis mentioned as examples "The heat was stifling" and "She rummaged in her handbag." Neither of these two specific phrases makes its way into the *Dictionary of Clichés*.

But beyond that, defining a cliché is time dependent. Some clichéd phrases such as "never darken my door again" predate even older authors like Jane Austen. Others are younger. "Yada yada" was repopularized by a 1997 *Seinfeld* episode and "soccer mom" has a clear modern connotation. "Catch-22," which Ammer includes, is based on the 1961 Joseph Heller novel. It wasn't cliché when Heller invented it. Language changes over time, and if Christine Ammer published this list 200 years earlier, in 1813 (the same year *Pride and Prejudice* was published), the results would look different.

Clichés That Authors Use in More Than Half Their Works

AUTHOR	WORKS	CLICHÉ
Isaac Asimov	7 Foundation Series books	*past history*
Jane Austen	6 novels	*with all my heart*
Enid Blyton	21 Famous Five books	*in a trice*
Ray Bradbury	11 novels	*at long last*
Ann Brashares	9 novels	*blah blah blah*
Dan Brown	4 Robert Langdon books	*full circle*
Tom Clancy	13 novels	*by a whisker*
Suzanne Collins	3 Hunger Games books	*put two and two together*
Clive Cussler	23 Dirk Pitt novels	*wishful thinking*
James Dashner	3 Maze Runner novels	*now or never*
Theodore Dreiser	8 novels	*thick and fast*
William Faulkner	19 novels	*sooner or later*
Dashiell Hammett	5 novels	*talk turkey*
Khaled Hosseini	3 novels	*nook and cranny*
E L James	3 Fifty Shades books	*words fail me*
James Joyce	3 novels	*from the sublime to the ridiculous*
George R. R. Martin	8 novels	*black as pitch*
Herman Melville	9 novels	*through and through*
Stephenie Meyer	4 Twilight books	*sigh of relief*
Vladimir Nabokov	8 novels	*in a word*
James Patterson	22 Alex Cross novels	*believe it or not*
Jodi Picoult	21 novels	*sixth sense*
Rick Riordan	5 Percy Jackson novels	*from head to toe*
J. K. Rowling	7 Harry Potter books	*dead of night*
Salman Rushdie	9 novels	*the last straw*
Alice Sebold	3 novels	*think twice*
Zadie Smith	4 novels	*evil eye*
Donna Tartt	3 novels	*too good to be true*
J. R. R. Tolkien	*LOTR* and *The Hobbit*	*nick of time*
Tom Wolfe	4 novels	*sinking feeling*

Note: These clichés are all from *The Dictionary of Clichés*. The selection in this table was chosen by me for variety and oddity, not by any specific quantitative measure.

The last question to consider, before crowning our champion of clichés, is whether to include clichés that appear in dialogue. If characters are using clichéd phrases, is it clichéd writing? Or just capturing how people talk? Take, for instance, this sample of dialogue from James Patterson's *Mary, Mary*, with phrases Ammer lists in her *Dictionary of Clichés* in bold. The main character, Alex Cross, is the first one to speak below.

> ". . . But if it gets you Mary Smith, then everything's okay and you're a hero."
> **"Russian roulette,"** she said dryly.
> **"Name of the game,"** I said.
> "By the way, I don't want to be a hero."
> **"Goes with the territory."**
> She finally smiled. "America's Sherlock Holmes. Didn't I read that somewhere about you?"
> "Don't believe everything you read."

The short dialogue above has three clichés from Ammer's set list of 4,000, as well as two other phrases ("I don't want to be a hero" and "don't believe everything you read") that would be strong contenders if Ammer ever expands to 4,002. And while one could argue against including dialogue, I've ultimately decided to leave it in. If so much of your main character's language is tied up in cliché, then at a certain point, so is your main character—and so may be your novel.

Now, with all these caveats out in the open, what do we find? Who is the most clichéd writer?

I've counted the total number of times any cliché appeared in an author's writing, looking at all fifty authors I've been using as examples throughout this book.

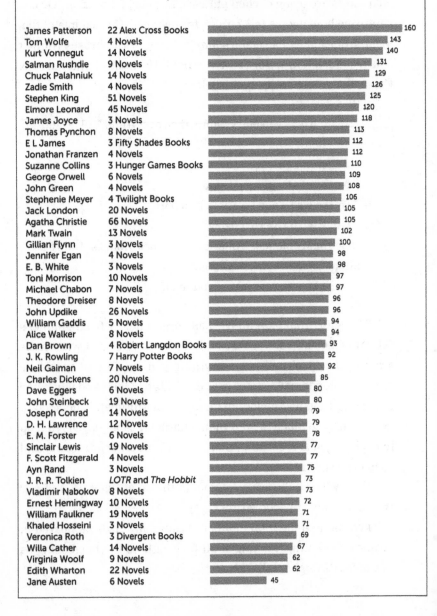

Number of Clichés per 100,000 Words

James Patterson	22 Alex Cross Books	160
Tom Wolfe	4 Novels	143
Kurt Vonnegut	14 Novels	140
Salman Rushdie	9 Novels	131
Chuck Palahniuk	14 Novels	129
Zadie Smith	4 Novels	126
Stephen King	51 Novels	125
Elmore Leonard	45 Novels	120
James Joyce	3 Novels	118
Thomas Pynchon	8 Novels	113
E L James	3 Fifty Shades Books	112
Jonathan Franzen	4 Novels	112
Suzanne Collins	3 Hunger Games Books	110
George Orwell	6 Novels	109
John Green	4 Novels	108
Stephenie Meyer	4 Twilight Books	106
Jack London	20 Novels	105
Agatha Christie	66 Novels	105
Mark Twain	13 Novels	102
Gillian Flynn	3 Novels	100
Jennifer Egan	4 Novels	98
E. B. White	3 Novels	98
Toni Morrison	10 Novels	97
Michael Chabon	7 Novels	97
Theodore Dreiser	8 Novels	96
John Updike	26 Novels	96
William Gaddis	5 Novels	94
Alice Walker	8 Novels	94
Dan Brown	4 Robert Langdon Books	93
J. K. Rowling	7 Harry Potter Books	92
Neil Gaiman	7 Novels	92
Charles Dickens	20 Novels	85
Dave Eggers	6 Novels	80
John Steinbeck	19 Novels	80
Joseph Conrad	14 Novels	79
D. H. Lawrence	12 Novels	79
E. M. Forster	6 Novels	78
Sinclair Lewis	19 Novels	77
F. Scott Fitzgerald	4 Novels	77
Ayn Rand	3 Novels	75
J. R. R. Tolkien	*LOTR* and *The Hobbit*	73
Vladimir Nabokov	8 Novels	73
Ernest Hemingway	10 Novels	72
William Faulkner	19 Novels	71
Khaled Hosseini	3 Novels	71
Veronica Roth	3 Divergent Books	69
Willa Cather	14 Novels	67
Virginia Woolf	9 Novels	62
Edith Wharton	22 Novels	62
Jane Austen	6 Novels	45

Out of the fifty authors I use as examples, James Patterson is the most clichéd writer. That's what the numbers say. It's not just his dialogue, as in this graph, that accounts for this; his prose sections are dense with familiar phrases as well. Here's an excerpt from the first Alex Cross book, with the phrases Ammer considers clichés in bold:

> Michael Goldberg weighed **next to nothing** in his arms, which was exactly what he felt about him. Nothing. Then came the little princess, the **little pride and joy**, Maggie Rose Dunne. All the way from **La-la-land** originally.

If you go over that list above you might begin to have some sympathy for the blockbuster author. His competition is stacked against him. Jane Austen, who had the lowest total cliché use, was writing two centuries ago. Though Stephen King, Dan Brown, and E L James are all on the list, it might be unfair to judge Patterson against Cather, Wharton, or Faulkner. It's possible that specific cliché phrases are so unique to time and reading audience that any comparison to older literary authors is meaningless.

But further tests show that there is a real difference in how these clichéd phrases are used in literary and mass-market fiction, even when controlling for time period. Of all Pulitzer Prize winners between 2000 and 2016, the median cliché rate is 85 per 100,000. If we look at the top ten bestselling books per year between 2000 and 2016 according to *Publishers Weekly*, the median cliché rate is 118—almost 40% higher.

Anthony Doerr's *All the Light We Cannot See* and Viet Thanh Nguyen's *The Sympathizer* are the two most recent winners of the Pulitzer Prize for fiction as of this writing. Per 100,000 words, they contain 39 clichés and 78 clichés respectively. Patterson's *Unlucky 13* and *Truth or Die*, which were published within a month of the two award winners above, contain clichés at a rate of 149 and 183 per 100,000 words.

Even in his genre of super popular fiction James Patterson scores ahead. I took all 127 novels that ranked as *Publishers Weekly* bestselling books of the year, going all the way back to 2000. This list features some breakout hits such as *The Help* and *The Lovely Bones*, but the majority are novels by authors in the same league as Patterson, including Nicholas Sparks, David Baldacci, Stephenie Meyer, J. K. Rowling, Suzanne Collins, John Grisham, Patricia Cornwell, and Tom Clancy. Of all these 127 books, the most popular of the twenty-first century, Patterson's seventeenth Alex Cross book, *Cross Fire*, has the most clichéd phrases. It contains a whopping 242 clichés per 100,000 words. And of the top five most clichéd books, four are by Patterson.

Most Clichéd Popular Books of the Twenty-First Century		
AUTHOR	BOOK	CLICHÉS PER 100,000 WORDS
James Patterson	*Cross Fire*	242
James Patterson	*Mary, Mary*	218
Jan Karon	*Light from Heaven*	218
James Patterson	*The Quickie*	215
James Patterson	*I, Alex Cross*	208
Janet Evanovich	*Fearless Fourteen*	206
James Patterson	*Kill Alex Cross*	204
Janet Evanovich	*Finger Lickin' Fifteen*	199
Janet Evanovich	*Plum Lovin'*	199
Tom Clancy	*Dead or Alive*	197

Patterson averages more than one of the top ten bestselling books per year. That speaks to his impressive dominance of the book market. But for him to take up half the spots on the list above speaks to his impressive use of the common cliché.

If you've read Patterson, his ranking is probably not surprising. And even if you've read just Patterson book titles, it should make sense. Patterson has novels titled *11th Hour*, *Cat & Mouse*, and *7th Heaven*, all of which are considered clichés in *The Dictionary of Clichés*. If we move beyond that master list, Patterson has invoked common phrases in his titles ranging from *Roses Are Red* to *Judge & Jury*, *Treasure Hunters*, *Unlucky 13*, *Sundays at Tiffany's*, and *First Love*.

Joseph Heller named his book *Catch-22*, which was so original and memorable that people started to copy him until it became cliché. Shakespeare coined phrases such as "all that glitters is not gold," "dead as a doornail," "heart of gold," "in a pickle," and "wild-goose chase." His words rose to the level of cliché when people integrated them into their own writing or speech. Patterson is the opposite. He integrates clichés into his own writing, but clearly, from his massive popularity, he's a master of it in his own way.

Would Amis agree with the numbers that Patterson is the most clichéd writer? As mentioned above, the two phrases Amis highlighted in his Charlie Rose interview as giveaways of clichéd writing were "The heat was stifling" and "She rummaged in her handbag." I went through my sample of fifty authors to see how many writers described heat as stifling and wrote "She rummaged in her handbag" in a book verbatim. There was one author who did both: James Patterson.

Sting Like a Simile

In his acclaimed novel *The Kite Runner* Khaled Hosseini offers a defense of clichés.

A creative writing teacher at San Jose State used to say about clichés: "Avoid them like the plague." Then he'd laugh at his own joke. The class laughed along with him, but I always thought clichés got a bum rap. Because, often, they're dead-on. But the aptness of the clichéd saying is overshadowed by the nature of the saying as a cliché. For example, the **"elephant in the room"** saying. Nothing could more correctly describe the initial moments of my reunion with Rahim Khan.

Writers and teachers warn of clichéd expressions that distract the reader. But as Hosseini points out with his animal comparison, sometimes the "elephant in the room" just captures the feeling you wish to describe. And perhaps that's why certain phrases get repeated *until* they're cliché.

Quiet as a mouse. Sly as a fox. Float like a butterfly. Sting like a bee. One of the most basic tools that writers have used throughout literary history, across time and culture and distance, is animal imagery—to the point where many of our comparisons to animals now count as cliché. Yet animal similes can still be potent and evocative. And part of a distinctive style.

While it won't catch every instance, I searched through a range of authors' animal similes using the following simple structure.

_____ like/as a(n) [optional adjective] [animal]

The formulation may seem so corny that authors would avoid it, but nearly all writers, popular and literary, find a use for animal similes. Of contemporary novelists, Stephen King is the third most prolific user, with 11 animal similes per 100,000 words, while Gillian Flynn and Neil Gaiman both measure in at 16.

The only author in my sample that I found never using such a primal simile is Jane Austen. Classics from the same time period (*Frankenstein, Ivanhoe, The Last of the Mohicans*) all have animal similes, so it's not just an artifact from her time period. She avoided them in all of her six books.

Some authors run the opposite. After all my searching, the novelist who used animal similes the most was D. H. Lawrence, author of *Lady Chatterley's Lover* and *Sons and Lovers*, among others.

Here is Lawrence compared to a slew of other prominent writers born within the same time period (1850–1899) as he was. Other than Steinbeck and Faulkner, he doubles everyone else in usage, including Jack London, known for his nature-heavy works.

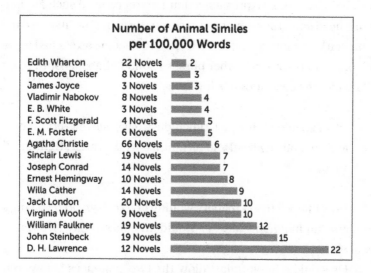

Number of Animal Similes per 100,000 Words		
Edith Wharton	22 Novels	2
Theodore Dreiser	8 Novels	3
James Joyce	3 Novels	3
Vladimir Nabokov	8 Novels	4
E. B. White	3 Novels	4
F. Scott Fitzgerald	4 Novels	5
E. M. Forster	6 Novels	5
Agatha Christie	66 Novels	6
Sinclair Lewis	19 Novels	7
Joseph Conrad	14 Novels	7
Ernest Hemingway	10 Novels	8
Willa Cather	14 Novels	9
Jack London	20 Novels	10
Virginia Woolf	9 Novels	10
William Faulkner	19 Novels	12
John Steinbeck	19 Novels	15
D. H. Lawrence	12 Novels	22

If you want a better sense of the breakdown of Lawrence's purple prose consider his love of birds. Or at least his love of bird imagery. The English novelist's first book was titled *The White Peacock*. His last book was *The Escaped Cock*. He uses comparisons to birds more than authors like Agatha Christie, F. Scott Fitzgerald, or James Joyce used *any* comparisons to animals. I counted 116 separate bird similes in his 12 novels.

Digging deeper, there are a few explanations for Lawrence's bird tic. D. H. Lawrence was a poet as well, and his best-known collection is *Birds, Beasts and Flowers*. As you can imagine, Lawrence's poetry about birds and beasts is very simile heavy. He exceeds a rate of 90 animal similes per 100,000 words. Furthermore, Lawrence was a travel writer known for his anti-industrial views. He was always drawn to nature. In his more philosophical nonfiction he called out moderners for losing touch with the physical world.

At the time, his sensory prose served him very well and his books are still popular among literary scholars. But it's hard to read some of his sentences in the twenty-first century and understand them in the same way they might have worked on readers 100 years ago. Expressions that Lawrence used such as "bury your head sometimes, like an ostrich in the sand" or "like a scared chicken" are understood, even if they've become as clichéd as "an elephant in the room." Other phrases sound almost absurd today. Consider the description below.

> "It's ridiculous. It's just ridiculous!" she blurted, bridling and ducking her head and turning aside, **like an indignant turkey**.

How many current readers wouldn't be thrown off by trying to imagine an indignant turkey? Would an editor today give this an axe without second thought?

Or would a book today allow the two sentences below to be printed on the same page? They both use the same hawk simile to describe eyes.

> "I don't know," he said, looking at his uncle with his bright inhuman eyes, **like a hawk's**.
>
> Will Brangwen ducked his head and looked at his uncle with swift, mistrustful eyes, **like a caged hawk**.

What's Your Favorite Word?

In the 1995 book *The Logophile's Orgy*, Lewis Burke Frumkes asked notable writers, including Ray Bradbury, to submit their favorite word. According to the *Fahrenheit 451* author:

> My two words are "ramshackle" and "cinnamon."

Bradbury gives an intricate reason for loving *ramshackle*:

> It's hard to explain why "ramshackle" has played such a part
> in my writing. . . . Half the time we feel we are ramshackle
> people, lopsided, no right or left side of the brain, with some
> terrible vacuum in between. That, to me, is ramshackle.

But his affinity for the word *cinnamon* is based on a more
deep-seated, personal connection:

> The word "cinnamon" derives, I suppose, from visiting my
> grandma's pantry when I was a kid. I loved to read the labels
> on spice boxes; curries from far places in India and cinna-
> mons from across the world.

Bradbury was deciding upon which word was his "favorite"
when he chose *ramshackle* and *cinnamon*. And if we look at the
numbers, he does indeed use these words more often than other
writers. Of the fifty authors I've used throughout this book as examples, ranging from J. K. Rowling to Vladimir Nabokov and Agatha Christie to Jane Austen, none used *ramshackle* as frequently
as Bradbury. And just one of the fifty, Toni Morrison, used *cinnamon* more often.

In his ode to *cinnamon*, Bradbury mentions that it reminds
him of reading the labels on spice boxes in his grandma's pantry.
Without reading Bradbury's explanation, would it have been possible, with a bit of statistical sleuthing, to detect this memory?

Of the 50 other authors in the sample, Bradbury uses *spice*
more than 48 of the other authors. His use of the word *curry*,
which he mentioned alongside cinnamon in his explanation, is not
high. However, many other flavors that might be associated with
a pantry, such as spearmint, vanilla, peppermint, nutmeg, onion,
licorice, and lemon, are used by Bradbury at a high rate in many of

his books. In several cases he uses the word more often than any of his peers. Below are the fifty authors and Bradbury ranked in terms of their use of each word. (No modification was made to the context of these flavor words. For instance, E L James's high use of *vanilla* in *Fifty Shades of Grey* can be explained by her use of the phrase "vanilla sex." *Curry* refers to the spice as well as the dish. The use of *curry* as a verb does not alter the rankings.)

	1ST	2ND	3RD
Cinnamon	Toni Morrison	Ray Bradbury	Khaled Hosseini
Curry	Neil Gaiman	Zadie Smith	J. K. Rowling
Spearmint	Ray Bradbury	Gillian Flynn	George Orwell
Spice	Salman Rushdie	James Joyce	Ray Bradbury
Vanilla	E L James	Ray Bradbury	Chuck Palahniuk
Peppermint	E. B. White	Ray Bradbury	George Orwell
Nutmeg	Ray Bradbury	Neil Gaiman	Chuck Palahniuk
Licorice	Ray Bradbury	Chuck Palahniuk	Jonathan Franzen
Onion	James Joyce	Michael Chabon	Ray Bradbury
Lemon	Ray Bradbury	Sinclair Lewis	James Joyce

Bradbury uses many of these spice words at an unusual level. If you put full faith in the data, you might even come to the conclusion that *cinnamon* is not his favorite word when compared to his fondness for the words *spearmint, vanilla, peppermint, licorice,* or *nutmeg.*

To narrow in on one word, his use of *spearmint* is much more anomalous than *cinnamon.* It's not surprising that Bradbury used *cinnamon* often; it's a common word. But no other author in my sample comes anywhere near Bradbury's use of *spearmint.* He uses the word nearly as many times in his 11 novels as all other fifty authors combined in their 550-plus books (including variants such as *spear-mint* or *spear mint*). Bradbury's love of *spearmint* is a true outlier. While responsible for more than 50% of all *spearmint* in the sample, his use of *cinnamon* accounts for just 6%.

Perhaps, out of curiosity, you want to see a larger sample than the fifty I selected. One safe option would be the Corpus of Historical American English. It's a sample of works from 1810 to 2009 that totals 385 million words (about six times the number of words in the sample of fifty authors) assembled by linguists at Brigham Young University. The corpus is not so much about creating a sample of notable novels (as my sample is). It's good for creating a benchmark for more ordinary writing over the last 200 years. In fact, in addition to fiction it also includes nonfiction books along with an assortment of magazine and newspaper articles. It's exemplary of written English as a whole over the last two centuries.

Ray Bradbury uses the word *cinnamon* 4.5 times more often than the word is used in the Corpus of Historical American English. Compared to ordinary writing, he does use the word often. But Bradbury uses *spearmint* a full fifty times more often than it's used in the same historical corpus. *Cinnamon* is Bradbury's "favorite" word, but *spearmint*, whether he realizes it or not, must clock in somewhere near the top.

Bradbury's use of spice words as a whole is high, but it's not so high as to be obtrusive or all that distracting. On average, it amounts to no more than one or two uses of each spice per book.

On the other hand, there can be tic words that writers end up leaning on *a lot*, to the point where they appear hundreds of times in a book and can even disrupt the reading experience. For instance, in the same collection of favorite words, Michael Connelly contributed his favorite as *nodded*. Connelly has written seven number one bestsellers and seen two of his movies be adapted into major films (*Blood Work*, starring Clint Eastwood, and *The Lincoln Lawyer*, starring Matthew McConaughey). He said this of his favorite word and his main character Harry Bosch:

> He's a man of few words. He reacts by nodding, so "nodding"
> ends up in all my books. I had an editor who pointed out

that Harry nods too much. In fact in one book he nodded 243 times.

Connelly used *nodded* (or the variants *nod, nods, nodding*) more than any of the fifty authors in the sample. He uses it 109 times per 100,000 words (about once every three or four pages). That's twice as often as Agatha Christie used it, four times as often as J. K. Rowling used it, and eight times as often as Ernest Hemingway. He uses *nodded* fifteen times more often than it's seen in the Corpus of Historical American English.

There are 119 words that Connelly uses at least once per thousand words. These are words like *looked, car, case, something, phone,* or *about.* Of these 119 words, the one word that Connelly uses most, relative to the Corpus of Historical American English, is *nodded.* Considering he didn't run any of the numbers but still came up with the number one word he uses disproportionally, it's fair to say Connelly has a great pulse on his own eccentricities.

If Connelly knows he uses *nodded* often, is that a bad thing? There's a fine line between a word you like and a word you're finding yourself relying on too heavily. As journalist Ben Yagoda writes in his book *How to Not Write Bad*:

Word repetition is a telltale—maybe the telltale—sign of awkward, nonmindful writing.

Yagoda goes on to explain in more detail the case against repeating words. He suggests not using the same word twice in a sentence, though he says there are some exceptions if it's a common enough word. With some words, which are distinct enough, he advises waiting several pages before returning to them.

It seems that for Connelly *nod* tends to fall into this latter category: It's a distinct enough word that Connelly uses often, but which most other writers tend to use more rarely. Connelly knows he likes to use it, but even so, there are times when its use seems almost automatic. Take these two (albeit some of the more extreme) examples.

From *Chasing the Dime*:

> "But we can go smaller," he said. "A lot smaller."
> She **nodded** but he couldn't tell if she saw the light or was just **nodding**.
> "Molecules," she said.
> He **nodded**.

And this exchange from *Lost Light*:

> "White guys?"
> I **nodded**.
> "Damn. That was good."
> I **nodded** again.
> "So what's under the tablecloth, Harry?"
> I shrugged.
> "First time you've come around in eight months, I suppose you know."
> She **nodded**.
> "Yeah."
> "Let me guess. Alexander Taylor's tight with the chief or the mayor or both and he called to check me out."
> She **nodded**. I had gotten it right.

I've come to think of words like this, which make their way into heavy rotation for particular authors, as their "fallback" words. They're favorites that go beyond rarity and become common for that author, almost as if they're a part of the way that author thinks and operates.

We all have our favorites and our fallbacks, our *cinnamon*

words and our *nod* words. I was curious: What do the numbers tell us about other authors' favorite and fallback words? I decided to make a loose set of rules to pinpoint these two categories, and at the end of the chapter I've scored the most extreme peculiarities for dozens of famous authors.

First, to find their favorites, their "cinnamon words," I've used the following set of requirements:

- It must be used be in **half** an author's books.
- It must be used at a rate of at least **once** per 100,000 words throughout an author's books.
- It must not be so obscure that it's used less than once per million in the Corpus of Historical American English.
- It is not a proper noun.

For each author, I found all words that passed the rules above and then homed in on the three that have the highest usage rate compared to the Corpus of Historical American English. These are the "cinnamon words," an approximation of their favorites.

For example, consider the list that we find when we look at Vladimir Nabokov's work. The *Lolita* author's favorite—his number one "cinnamon word," used at least once in all of his eight books—is *mauve*. In total, he used it at a rate 44 times more common than the word is used in the Corpus of Historical American English. No other word in Nabokov's work shows such a big difference when compared to ordinary writing.

And it makes perfect sense that a color, such as mauve, would be one of Nabokov's "cinnamon words." He was known to have synesthesia. Or, as he describes, with plenty of detail and color (including mauve), in his autobiography *Speak, Memory*:

> At times, however, my photisms take on a rather soothing *flou* quality, and then I see—projected, as it were, upon the

inside of the eyelid—gray figures walking between bee-hives, or small black parrots gradually vanishing among mountain snows, or a **mauve** remoteness melting beyond moving masts.

On top of all this I present a fine case of colored hear-ing. Perhaps "hearing" is not quite accurate, since the color sensation seems to be produced by the very act of my orally forming a given letter while I imagine its outline. The long *a* of the English alphabet (and it is this alphabet I have in mind farther on unless otherwise stated) has for me the tint of weathered wood, but a French *a* evokes polished ebony.

Given the unique way Nabokov described his thoughts, it seems as if the "cinnamon word" method was able to succeed in landing upon a word that was unique to him. His love of mauve is extraordinary, but he uses all colors more than other writers as well. If you use the 64 standard Crayola Crayon names as a defini-tive list of colors, Vladimir Nabokov uses around 460 color words per 100,000, which is remarkably high. The same colors appear just 115 times per 100,000 in the Corpus of Historical American English.

Not everyone's "cinnamon words" may be as telling as Nabokov's, and the method isn't perfect. For a number of authors, the words reflect the unique tone of a book or subject matter. Jane Austen's top three are *civility*, *fancying*, and *imprudence*, while Agatha Christie's are *inquest*, *alibi*, and *frightful*. For authors like J. K. Rowling, where I chose to include just the one series for which she is best known, the words are representative of that universe itself rather than any words representing a likely favorite. The top three "cinnamon words" for the Potter series are *wand*, *wizard*, and *potion*. For *Fifty Shades* they're *murmurs*, *hmm*, and *subcon-scious*. And for Patterson's Alex Cross books we get *killers*, *mur-ders*, and *kidnapping*—which could function well as a tagline.

I've also developed a set of rules for finding each author's fall-backs, their "nod" words, which an author uses over and over again to the point where it gets noticeable. I've defined a "nod" word by the following requirements:

- It must be in **all** of an author's books.
- It must be used at a rate of at least **100** per 100,000 words throughout an author's books.
- It must not be so obscure that it's used less than once per million in the Corpus of Historical American English.
- It is not a proper noun.

The top three "nod" words were then calculated in the same way, by comparing usage rates to the Corpus of Historical American English. These words too are sometimes taken over by subject and setting (Suzanne Collins's fallback words of the Hunger Games series include *district* and *games*), and they tend to be drier, blander words. But they can also be revealing of the inner mechanics of certain authors' writing—the devices and tics they tend to fall back on to keep the plot moving or to get from one scene to the next. Gaiman fills the gaps with *walking*; Cheever's reality is shifting, focusing on how things *seem*; Stine's Goosebumps books are filled with *staring* and *crying*; some authors focus on *feel*, others on *want*.

Following is a chart of the top three "cinnamon words" and top three "nod words" for each of the fifty authors I've been using as examples throughout this book. To add to the fun, I've also included fifty more authors of popular and critical acclaim. Most authors' words don't reveal a deep truth—though a good number offer brief glimpses of how these authors work and think compared to one another.

AUTHOR	WORKS	CINNAMON WORDS	NOD WORDS
Chinua Achebe	5 Novels	kinsmen, abomination, compound	girls, room, likes
Douglas Adams	7 Novels	galactic, spaceship, robot	yes, said, just
Mitch Albom	6 Books	exhaled, hmm, mumbled	phone, felt, asked
Isaac Asimov	7 Foundation Series Books	galactic, terminus, councilman	second, said, yes
Jean Auel	6 Earth's Children Books	totem, clan, steppes	clan, cave, wolf
Jane Austen	6 Novels	civility, fancying, imprudence	herself, dear, lady
David Baldacci	29 Novels	web, laptop, limo	looked, really, back
Enid Blyton	21 Famous Five Books	woof, hallo, larder	dog, round, said
Ray Bradbury	11 Novels	icebox, dammit, exhaled	someone, cried, boys
Ann Brashares	9 Novels	smock, dorm, boyfriend	maybe, felt, herself
Charlotte Brontë	4 Novels	tradesman, gig, lineaments	my, am, me
Dan Brown	4 Robert Langdon Books	grail, masonic, pyramid	felt, toward, looked
Truman Capote	5 Novels	geranium, icebox, drugstore	though, liked, seemed
Willa Cather	14 Novels	cottonwood, hearted, lilac	went, always, looked
Michael Chabon	7 Novels	nostalgia, boardwalk, fucked	black, around, said
John Cheever	5 Novels	infirmary, venereal, erotic	seemed, went, asked
Agatha Christie	66 Novels	inquest, alibi, frightful	yes, quite, really
Tom Clancy	13 Novels	briefed, sniper, gunmen	sir, asked, something
Cassandra Clare	9 Novels	inquisitor, vampire, demons	blood, hair, looked
Suzanne Collins	3 Hunger Games Books	tributes, tracker, victors	district, games, says
Michael Connelly	27 Novels	freeway, homicide, laptop	nodded, phone, car
Joseph Conrad	14 Novels	immobility, poop, skylight	seemed, voice, head
Michael Crichton	24 Novels	dinosaur, sensors, syringe	said, yes, looked
Clive Cussler	23 Dirk Pitt Novels	underwater, hangar, artifact	ship, sea, water

AUTHOR	WORKS	CINNAMON WORDS	NOD WORDS
James Dashner	3 Maze Runner Novels	glade, flare, dude	finally, maybe, felt
Don DeLillo	15 Novels	tempo, era, carton	off, said, come
Charles Dickens	20 Novels	rejoined, waistcoat, workhouse	sir, dear, am
Theodore Dreiser	8 Novels	genially, franchises, subtlety	anything, oh, might
Jennifer Egan	4 Novels	blah, backpack, glimpsed	felt, looked, eyes
Dave Eggers	6 Novels	kayak, watchers, laptop	wanted, hand, knew
Jeffrey Eugenides	3 Novels	manic, backseat, lifeboat	girls, room, like
Janet Evanovich	40 Novels	stun, backseat, doughnut	car, lot, maybe
William Faulkner	19 Novels	hollering, realized, immobile	maybe, even, already
Joshua Ferris	3 Novels	website, totem, bookshelves	office, asked, wanted
F. Scott Fitzgerald	4 Novels	facetious, muddled, sanitarium	oh, seemed, night
Ian Fleming	12 James Bond Novels	lavatory, trouser, spangled	round, across, girl
Gillian Flynn	3 Novels	fucked, pissed, fridge	hair, girl, really
E. M. Forster	6 Novels	muddle, hullo, tram	oh, yes, she
Jonathan Franzen	4 Novels	carpeting, earthquakes, dorm	want, she, her
Charles Frazier	3 Novels	poplar, forearms, cove	fire, dark, ground
William Gaddis	5 Novels	suing, damned, someplace	damn, mean, wait
Neil Gaiman	7 Novels	unimpressed, glinted, eyebrow	walked, door, said
Mark Greaney	6 Novels	backpack, surveillance, gunfire	court, front, behind
John Green	4 Novels	prom, pee, backpack	yeah, maybe, really
John Grisham	28 Novels	paperwork, courtroom, juror	office, asked, money
Dashiell Hammett	5 Novels	coppers, taxicab, sidewise	asked, anything, got
Nathaniel Hawthorne	6 Novels	subtile, betwixt, remoteness	heart, seemed, might
Ernest Hemingway	10 Novels	concierge, astern, cognac	said, big, asked
Khaled Hosseini	3 Novels	kites, backseat, orphanage	father, eyes, around

AUTHOR	WORKS	CINNAMON WORDS	NOD WORDS
E L James	3 Fifty Shades Books	murmurs, hmm, subconscious	murmurs, fingers, mouth
Henry James	20 Novels	recognise, oddity, afresh	herself, mean, moment
Edward P. Jones	3 Novels	coop, heh, icebox	street, woman, children
James Joyce	3 Novels	tram, bello, hee	old, your, his
Stephen King	51 Novels	goddam, blah, fucking	looked, back, around
Rudyard Kipling	3 Novels	job, hove, camel	thee, till, work
D. H. Lawrence	12 Novels	tram, realized, sheaves	round, dark, sat
Elmore Leonard	45 Novels	fucking, shit, bullshit	saying, looking, said
Ira Levin	7 Novels	foyer, snowflakes, carton	smiled, said, looked
C. S. Lewis	7 Narnia Books	dwarfs, witch, lion	lion, king, round
Sinclair Lewis	19 Novels	golly, heh, darn	oh, room, going
Jack London	20 Novels	snarl, bristled, unafraid	knew, head, eyes
Lois Lowry	4 Giver Books	nurturing, mentor, seer	nodded, felt, told
George R. R. Martin	8 Novels	dragons, cloaks, whores	lady, red, black
Cormac McCarthy	10 Novels	yessir, mam, upriver	horses, watched, road
Ian McEwan	13 Novels	lavatory, forwards, fridge	room, hand, took
Richelle Mead	23 Novels	guardians, vampire, dorm	really, wanted, me
Herman Melville	9 Novels	whale, forecastle, sperm	sea, upon, though
Stephenie Meyer	4 Twilight Books	vampire, grimaced, flinched	voice, my, eyes
David Mitchell	6 Novels	mam, dint, piss	my, says, your
Toni Morrison	10 Novels	messed, navel, slop	she, women, her
Vladimir Nabokov	8 Novels	mauve, banal, pun	black, my, old
George Orwell	6 Novels	beastly, quid, workhouse	round, kind, money
Chuck Palahniuk	14 Novels	fingernail, backseat, orgasm	says, inside, dead
James Patterson	22 Alex Cross Novels	killers, murders, kidnapping	maybe, asked, right
Jodi Picoult	21 Novels	courtroom, diaper, diner	says, my, going
Thomas Pynchon	8 Novels	someplace, paranoia, freeway	here, around, back
Ayn Rand	3 Novels	transcontinental, comrade, proletarian	stood, felt, voice
Rick Riordan	5 Percy Jackson Novels	campers, titans, monsters	camp, looked, half

AUTHOR	WORKS	CINNAMON WORDS	NOD WORDS
Marilynne Robinson	4 Novels	checkers, baptized, pancakes	laughed, father, child
Veronica Roth	3 Divergent Books	simulation, serum, faction	says, gun, walk
J. K. Rowling	7 Harry Potter Books	wand, wizard, potion	wand, lit, professor
Salman Rushdie	9 Novels	whores, unleashed, fucking	love, her, too
Alice Sebold	3 Novels	dorm, rape, virginity	inside, father, my
Zadie Smith	4 Novels	fag, nah, backside	really, just, oh
Lemony Snicket	13 Unfortunate Events Books	siblings, orphans, villainous	siblings, orphans, children
Nicholas Sparks	18 Novels	peeked, owed, adrenaline	final, wanted, real
John Steinbeck	19 Novels	inspected, squatted, rabbits	got, looked, said
R.L. Stine	62 Goose-bumps Books	sneakers, whoa, creepy	backpack, stared, cried
Amy Tan	6 Novels	noodles, courtyard, leftover	my, told, saw
Donna Tartt	3 Novels	rimmed, ashtray, creepy	looking, around, said
J. R. R. Tolkien	*LOTR* and *The Hobbit*	elves, goblins, wizards	ring, dark, road
Mark Twain	13 Novels	hearted, shucks, satan	got, thing, yes
John Updike	26 Novels	rimmed, prick, fucked	like, her, face
Kurt Vonnegut	14 Novels	limousine, incidentally, foyer	said, war, father
Alice Walker	8 Novels	naw, fucked, lovemaking	black, white, women
Edith Wharton	22 Novels	nearness, daresay, compunction	herself, seemed, her
E. B. White	3 Novels	storekeeper, boatman, gander	replied, asked, heard
Tom Wolfe	4 Novels	fucking, haw, goddamned	black, looked, toward
Virginia Woolf	9 Novels	blotting, mantelpiece, diary	herself, she, looking
Markus Zusak	5 Novels	fellas, nah, footpath	street, words, girl

Chapter 8

How to Judge a Book by Its Cover

I wanted to see my name on the cover of a book. If your
name is in the Library of Congress, you're immortal!

—TOM CLANCY

Y ou know the old cliché by now: You shouldn't judge a book by
its cover. That doesn't mean you *can't* though. There are certain things you can glean from just a passing glance at the cover.
The author's stature, for instance.

Consider the case of Stephen King. His first book, *Carrie*, was
published in 1974. It never reached the top of the *New York Times*
bestseller list but it was an instant hit. It sold more than a million
copies in its first two years* and was adapted into a movie soon
after. When people bought *Carrie*, they were motivated by the
word of mouth alone. They had no preconception of Stephen King,
the author. Take a look at the first edition cover on the next page.

---

* According to a 2013 *Mental Floss* article, the paperback version, released a
year after the hardcover, was responsible for most of the huge sales volume.

King's name is small. It doesn't take up much of the whole cover, and it's modest in comparison to the title. In total, the author's name takes up less than 3% of the entire cover. (That is, if you were to draw a box around the name as tight as possible and compare that area to the entire cover, it would be less than 3%.)

"Stephen King" is not what your eyes are drawn to, but this is the last time his name could ever be missed. King's reputation grew, and so did

the size of his name. In all books after *Carrie* his name was larger and in some cases the growth is huge. For *Salem's Lot*, King's second book, his name reached 7% of the cover. In 1989, it reached an all-time high with his book *The Dark Half*, where a full 47% of the cover was devoted to King's name. Of the several hundred books I measured for this chapter, *The Dark Half* had the largest author's name of all.

The book's title is almost an afterthought. Readers are encouraged to buy it based on the writer's reputation alone.

This was exactly the fate King was trying to avoid when he created the pseudonym Richard Bachman. King wrote five books under the Bachman name before being outed as the author in 1984. On the opposite page we can see how differently Bachman's name was marketed than King's. Each book is charted

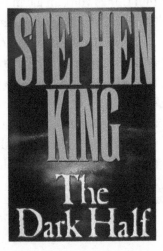

by the size of the author's name on the cover. As you can see, the six books with the smallest name are the five Bachman books and *Carrie*, which King wrote before he was famous.

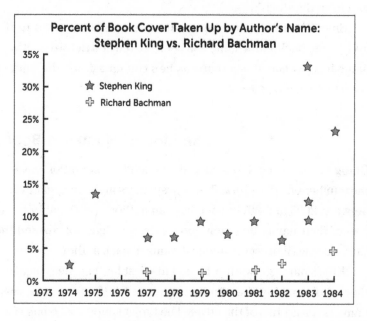

While there isn't a wealth of data points, I think that there is a clear trend to be discerned from the chart above. The fact that King's first book was the one where his name was the smallest reveals how book marketers think. If you haven't sold any books, your name is not a selling point, so the publisher keeps it small. If you've already had a bestseller and your name would bring those book buyers back around, then they might as well put that name in large print. We see the same thing with Bachman: Only as the Bachman name built into a (smallish) brand did the name size increase.

Keeping the Bachman name small is also in line with how King described his pseudonymous undertaking. In the introduction to *The Bachman Books*, which was published after his identity was revealed, King wrote: "The Bachman novels were 'just plain books,' paperbacks to fill the drugstore and bus-station racks of America. This was at my request; I wanted Bachman to keep a low profile. So, in that sense, the poor guy had the dice loaded against him from the start."

Since the Bachman-King connection was discovered in the mid-1980s, Bachman's name has been retired and the size of King's name has plateaued—perhaps as he's run up against the limit of good taste in cover design.

The Biggest Names in Books

Going beyond King, I looked at every single first edition cover of the number one *New York Times* bestsellers in the ten-year period between 2005 and 2014 to see how an author's profile affects the size of their name. For each book, I drew a tight box around the first, middle (if present), and last name of each author.

King's name grew after he had his first hit. For the typical author who's making the number one spot for the first time, their name takes up 12% of the cover. The typical author who has risen to become a bestseller machine, say with five-plus number one bestsellers, takes up 20% of the cover. By the time an author's name is well established, their name has almost doubled in size.

Opposite is a graph showing the typical size of the author's name. The center line is the median, and each box represents the middle 50%. (For instance, of all authors who had one previous bestseller the median name size was 18% of the cover and the middle 50% of all author names were between 12% and 23%.)

The pattern is there, even if far from absolute. The size of the name increases with more bestsellers, but only up to a point. At a certain level of fame, your name plateaus instead of growing and growing until it wraps around the cover from front to back.

By looking at the start of a bestselling author's career, it's easy to see the name increase by the time the authors transitions from

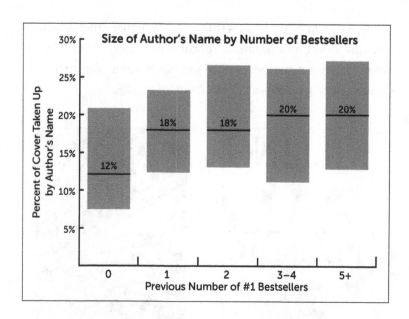

Size of Author's Name by Number of Bestsellers

Percent of Cover Taken Up by Author's Name (y-axis): 5%, 10%, 15%, 20%, 25%, 30%

Previous Number of #1 Bestsellers (x-axis): 0, 1, 2, 3–4, 5+

- 0: 12%
- 1: 18%
- 2: 18%
- 3–4: 20%
- 5+: 20%

nobody to star. There's an explosive growth from first book to repeat bestseller. Take Patricia Cornwell's first novel, in 1990, and her most recent number one bestseller (her ninth) below. Her name goes from 2% to 30%.

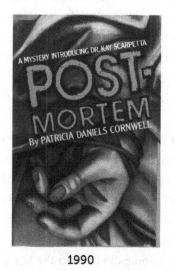

1990

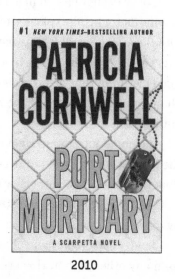

2010

Or Lee Child, who has written eight number one bestsellers. His first novel and his most recent number one bestseller are below. A quick glance shows the growth in name size, and a precise measuring reveals a growth from 5% to 22%.

1997

2014

Cover design has changed over the years, but that's not the driving factor in the explosion of Cornwell's or Child's name. King's *The Dark Half*, in which his name took up almost half the cover, was published in the 1980s, before Cornwell or Child started writing. One thing that has remained constant, even as cover styles change, is that the size of the author's name will grow with the author's sales.

Big-name authors are big-name brands, but at the extreme ends the biggest *name* does not necessarily lead to the *biggest* name. Following are all 26 authors with at least three bestsellers in the last ten years ranked by the size of their names. They are ranked by the typical (median) percent of the cover their name takes up.

James Patterson has sold more books in the last ten years than any other author. However, he comes in toward the bottom

The Biggest Names in Books		
AUTHOR	#1 *NYT* BESTSELLERS	MEDIAN SIZE OF NAME
Nora Roberts	7	37%
Harlan Coben	7	34%
J. D. Robb	8	34%
Mary Higgins Clark	9	32%
Patricia Cornwell	5	28%
Daniel Silva	6	27%
John Sandford	5	26%
Michael Connelly	7	26%
Janet Evanovich	15	24%
Vince Flynn	5	22%
Laurell Hamilton	4	22%
David Baldacci	13	22%
John Grisham	10	17%
J. R. Ward	3	17%
Lee Child	8	17%
Ken Follett	4	16%
Danielle Steel	3	16%
Stephen King	9	16%
Charlaine Harris	5	16%
Dean Koontz	5	14%
Jodi Picoult	7	13%
Sue Grafton	4	12%
James Patterson	10	12%
Jim Butcher	4	9%
Nicholas Sparks	8	5%
Mitch Albom	3	3%

of the chart. He lets his titles, often named to be a clear part of a series (*I, Alex Cross* or *Kill Alex Cross*) take a larger role than his own name. Stephen King has more fans than any other of the above authors on the book networking site Goodreads, but also has a reasonable-sized name on most of his covers. (Perhaps he's decided to dial things down after the 1980s.)

The top three authors are Nora Roberts, Harlan Coben, and

J. D. Robb. And it's important to note that J. D. Robb *is* Nora Roberts. Like King, Roberts created the pseudonym so she could publish more books without diluting her own brand, though this was never kept a secret like Bachman was. So if you want to know the true biggest name in books, there you have it—the Robb/Roberts goliath.

Close Quarters

If you ever take a few moments to look at the covers of new releases in popular fiction an interesting trend will stick out right away: Many books today are written by two authors. But if you haven't noticed, it might not be your fault. The size of the second author is often so small it might as well be invisible.

Even in the last couple of decades the practice of co-writing has become more common among blockbuster authors. In 1994 about 2% of all *New York Times* bestsellers were co-writes, while in 2014 10% were. There is no set procedure for listing co-authors and there is a wide range in how co-authors are presented. After all, the division of labor for co-authored books can be drastically different. Ghostwriters who write memoirs are not expecting to see their names on the cover. Neither would someone who writes the substance and conducts the research of a book for a politician. But in fiction, there's an expectation that those who write with others will include both names on the cover. You'll see "Tom Clancy **with** Peter Telep" or "James Patterson **and** Richard DiLallo." The meaning of the **with** and **and** may vary from author to author or book to book, but even the most mass-market writers are expected to give their co-writers some credit.

There are of course exceptions. On the opposite page is the cover for Glenn Beck's *The Christmas Sweater.* Though a Glenn Beck book, it is a full-length novel that reached the number one spot on the *New York Times* bestseller list. The book is listed almost everywhere, including on the *New York Times* list and on its Amazon page, as "Glenn Beck with Kevin Balfe and Jason Wright." But on the cover no space is saved for Balfe or Wright.

In another point of comparison, here's the cover of a book "co-authored" by Tom Clancy and Grant Blackwood.

Blackwood's name isn't all that noticeable next to Clancy's, but at least it's there. Clancy's name is 27 times larger than his co-author's. Many supporting writers, like Blackwood, are used to the treatment. Blackwood has never penned a bestseller himself but is the co-author with three different writers who all made the *New York Times* bestseller list. He is also the second author on both Clive Cussler and James Rollins thrillers, where his name is one-seventh and one-fourth the size of his co-authors'.

It is true that any of these big-name authors could share more of the page if they wanted to. Peter

Straub and Stephen King co-wrote two books together, *The Talisman* and *Black House*, over 15 years apart. Peter Straub is an acclaimed horror writer in his own right, but nowhere near the level of fame as Stephen King. I would guess the percentage of people reading *The Talisman* because of Straub instead of King would be relatively small. But on the two King-Straub covers, the names are the same size.

It's very rare, however, for authors of different statures to have

1990   2010

equal-sized names on the jacket. Clive Cussler penned six best-selling books with his own son, Dirk Cussler. Even on these books Clive Cussler's name is six times larger than his familial co-author. Likewise, British mystery writer Dick Francis has penned bestselling novels with his son Felix and his name is three times as large.

I've measured the ratio for all authors to have at least two co-authored books appear on the *New York Times* bestseller list between 2000 and 2014. Sometimes this can vary greatly between novels—James Patterson has some books where his name is twenty times larger than his co-author's, even though more often than not his name is within twice as large—but the results tend to average out.*

---

* In cases where writers work as a true pair, the first author listed on the cover was considered as the "author" in the chart—even if there is no other indication of him or her being the primary author.

Sharing the Spotlight: Size of Author's Name vs. Co-Author's

AUTHOR	HOW MANY TIMES LARGER PRINCIPAL AUTHOR'S NAME IS (MEDIAN)	CO-AUTHOR(S)
Glenn Beck	∞*	Jack Henderson, Harriet Parke, Kevin Balfe, Jason Wright
Oliver North	28X	Joe Musser
Tom Clancy	25X	Mark Greaney, Peter Telep, Grant Blackwood
Clive Cussler	12X	Graham Brown, Justin Scott, Jack Du Brul, Thomas Perry, Dirk Cussler, Grant Blackwood, Russell Blake, Paul Kemprecos
Catherine Coulter	8X	J. T. Ellison
W. E. B. Griffin	7X	William E. Butterworth IV
Rita Mae Brown	6X	Sneaky Pie Brown (Rita Mae Brown's "talking" cat)
Dick Francis	4X	Felix Francis
James Rollins	4X	Grant Blackwood, Rebecca Cantrell
Janet Evanovich	3X	Lee Goldberg
James Patterson	1.5X	Michael Lewidge, David Ellis, Maxine Paetro, Marshall Karp, Mark Sullivan, Emily Raymond, Howard Roughan, Richard Dilallo, Neil McMahon, Liza Marklund, Gabrielle Charbonnet, Andrew Gross, Peter de Jonge
Guillermo Del Toro	1.25X	Chuck Hogan
Brian Herbert	1X	Kevin Anderson
Douglas Preston	1X	Lincoln Child
Margaret Weis	1X	Tracy Hickman
Newt Gingrich	1X	William R. Forstchen
Tim LaHaye	1X	Jerry Jenkins, Bob Phillips

* Did not list co-authors on two of three books.

Bigger Books

When Edward Stratemeyer sat down and created the guidelines for the Hardy Boys he mandated not just that all books be the same number of chapters, but that they all have the same number of pages. He wanted books either 215 pages or 216 pages in length; 217 pages was too long, and 214 was too short. Over the years Leslie McFarlane would submit manuscripts to Stratemeyer that were on the short side and receive responses like these from the publishing magnate:

Can't you drill in some extra pages which I can insert

Please be sure to see to it that the story is the proper length to fill 216 text pages

They should be four to five pages longer, as I had to make additions in every instance to bring them up.

Stratemeyer's insistence on consistency was extreme and formulaic to the point where it was obtrusive. But, as someone who had seen how a series could shift or bloat as it progresses, he wanted to ensure there was not a dramatic change in the books' style or pace as the books kept coming.

Before Suzanne Collins wrote the smash-hit Hunger Games trilogy, she wrote a five-part young adult series called the Underland Chronicles. Those five books ranged from 57,000 to 79,000 words. It's not a wild difference, but judging by her future word counts Collins decided to take a page out of the Stratemeyer book when she started the Hunger Games.

Like Stratemeyer designing the Hardy Boys, Collins knew before the first book came out that she would be writing a series. The first Hunger Games was not published until 2008, but she signed a large advance for the entire trilogy in 2006. The three books measure in at near identical lengths of about 100,000, 102,000, and 101,000 words. And just like the Hardy Boys chapter mandate, all books in the Hunger Games are broken into three parts each consisting of nine chapters. The sole exception is a 350-word epilogue at the end of the final book. Otherwise, each book has the same organization down to the length of each section. The shortest part in any of the three books is 30,900 words, while the longest is 35,400.

Deciding upon such a strict structure, even before the first

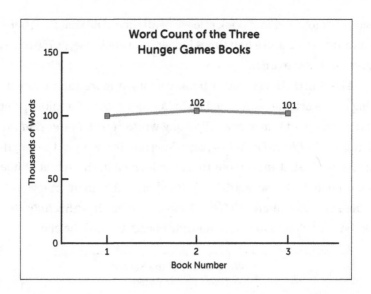

Word Count of the Three Hunger Games Books

book in a series has been published, may sound overbearing at first. But what happens when an author doesn't plan so far in advance? What happens when a new author finds himself with a best-seller on his hands and in need of a series?

Recent history suggests that establishing a Collins-like structure after the fact is a challenge. Consistency of length is rare. When authors find themselves in a bubble of popularity, book inflation is the norm.

The clearest example of this phenomenon is Rowling's Harry Potter series. The first book, *Harry Potter and the Sorcerer's Stone,* was released in 1997 and measures in at 309 pages (about 84,000 words). In 1997 interviews, after scoring a six-figure advance for the American rights, Rowling stated that she planned to write seven books. However, until the release of the first book she and her editor had no idea the books would achieve any level of success. The original advance for the first book was under $3,000. At the time J. K. Rowling was an unknown author and there was little fanfare to the event. Ten years later the final book, *Harry Potter and the Deathly Hallows,* was published in what was perhaps the most awaited book release of all time. And that final book was almost two and a half times longer than book one, at 759 pages (about 197,000 words).

The fourth Harry Potter book was itself more than twice as long as the original and ended up breaking a record for initial print run. Coming off that success Rowling wrote *Harry Potter and the Order of the Phoenix*. At 870 pages long and three times the length of the original, it turned out to be the longest in the series. While doing publicity for her sixth book, Rowling said in an interview with *Time*'s Lev Grossman that "I think *Phoenix* could have been shorter. I knew that, and I ran out of time and energy toward the end."

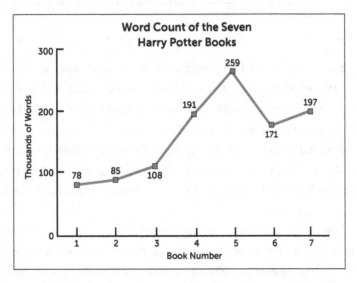

In similar situations, where an author with a hit finds themselves in a publishing pandemonium, the results tend to follow the same trajectory. Consider the Twilight series by Stephenie Meyer, *Fifty Shades of Grey* by E L James, and *Divergent* by Veronica Roth. Stephenie Meyer wrote on her blog that as she worked on Twilight she "wasn't planning a sequel" and Roth said she wrote the *Divergent* novel as "a stand-alone novel with series potential." Like Rowling, all three authors had never been published before their first book found a massive audience. And like Rowling, all three wrote longer and longer books as their series progressed.

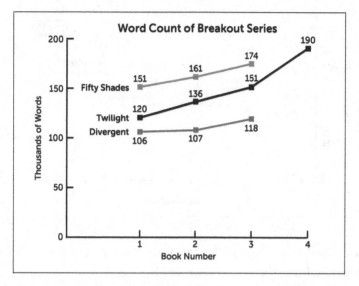

Word Count of Breakout Series

Unless there's a plan in place from the start, in each of these cases the books seem to inch ever bigger (unless, as Rowling did in book six, an author starts to fight back against their own book inflation). And when we look further, we find that it's not just blockbuster series where book inflation is apparent.

Amy Tan was 37 years old when her first novel, *The Joy Luck Club*, was published. Tan was not a known writer but *The Joy Luck Club* found immediate success in all ways. It was a finalist for the National Book Award and for the National Book Critics Circle Award. It also spent 32 weeks on the *New York Times* bestseller list. The novel, which is divided structurally into intersecting stories, is not long. It's 95,000 words and in its first edition was 288 pages.

Tan's next book, *The Kitchen God's Wife*, was 163,000 words. That's 70% longer. Tan has published five books since *The Joy Luck Club*, though none of them have quite replicated its critical acclaim and popularity. Her first and shortest book remains her

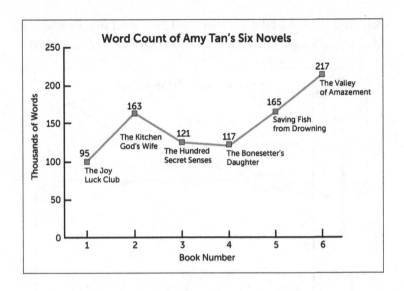

Word Count of Amy Tan's Six Novels

95 — The Joy Luck Club
163 — The Kitchen God's Wife
121 — The Hundred Secret Senses
117 — The Bonesetter's Daughter
165 — Saving Fish from Drowning
217 — The Valley of Amazement

(y-axis: Thousands of Words; x-axis: Book Number)

best known and best reviewed. Tan's most recent book is more than twice as long as her breakout hit.

In 1980, those who award the Pulitzer Prize for fiction started to announce finalists in addition to its winner. Since then, 24 writers have been fortunate enough to see their first novel receive immediate recognition as either a Pulitzer winner or finalist, or as a winner or finalist for the two prestigious awards that honored Tan: the National Book Award and the National Book Critics Circle Award.

Imagine being a writer who has spent years dreaming of writing a book, selling the novel to a publisher, and then having critics fall in love with it so much that they name it one of the best books of the year. It's a far-fetched fantasy, but for 24 authors since 1980 it's been a reality. How did these writers follow up their massive breakthrough success? Of these 24 first-time novelists, 17 came back with a second novel longer than their first. Seven came back with a shorter novel. It turns out that book inflation recurs throughout the literary world.

Twenty-four cases is not the biggest sample set, but if we want to examine just the case studies in which the author went from

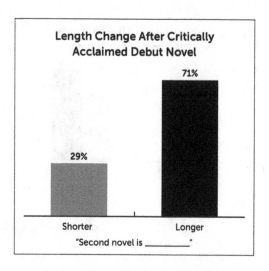

Length Change After Critically Acclaimed Debut Novel

71%

29%

Shorter   Longer
"Second novel is _____"

an unknown to a literary star overnight then we're constrained to study a small patch of anomalies. Even so, that 71% of these authors wrote longer books is a substantial trend.

The five authors who had their first novel nominated for a Pulitzer all came back with a second novel longer than their first. William Wharton's first novel, *Birdy*, was a finalist for the Pulitzer Prize in 1980. Wharton was over fifty years old when it came out, so it's not like he didn't have time to think it over. Still, after a half-century of not publishing anything, he came back three years later with a second novel, *Dad*, that was more than 40% longer.

It's worth noting that not every increase was as large as Tan's 70% leap or Wharton's 40%. If we break down the sample more, the results are perhaps a bit clearer. Below I've set a 20% change as the arbitrary barrier between the second novel being "Much Longer" or just "Slightly Shorter" (as well as the barrier between "Much Shorter" or "Slightly Shorter"). Forty-two percent of these breakout authors, almost half, wrote follow-up books that were "Much Longer." Only about one in ten authors wrote a follow-up that was "Much Shorter."

Book inflation is real, and the examples point to several pos-

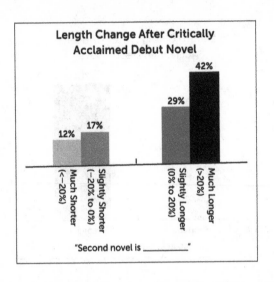

Length Change After Critically Acclaimed Debut Novel

42%

29%

17%

12%

Much Shorter
(<−20%)

Slightly Shorter
(−20% to 0%)

Slightly Longer
(0% to 20%)

Much Longer
(>20%)

"Second novel is _____"

sible explanations. When a book is as massive a success as Harry Potter there might be less incentive for the author or editor to aim for a shorter book. If the unknown Rowling had written an 870-page version of the first book in 1997, it would likely have had a much harder time getting published (and getting readers to pick it up). But after that first hit, once fans are invested and eager for more, any initial concerns about length become far less pressing. Indeed, by the last few Harry Potter books, many readers didn't want the adventures to end: more pages (i.e., more story) was *a blessing.*

When a writer is starting out, it's fair to say that there is a bottleneck against the length of books they can readily publish and sell. And if this is the case for a given author, then perhaps their first novel is not "short"; instead the second novel may be "reverting" to the author's natural book length. Likewise, if you are a breakout literary phenom lavished with praise, maybe you now have the chance to aim for a more epic and ambitious story on the next book than you could afford for your debut. It's also possible that writers who are nominated for prestigious awards on their first attempt feel they must top themselves to create a separate work of even greater import.

In the award-winning debut novel sample, it's impossible to answer if quality declined as length increased. In that sample of 24 writers, by definition, the first books were home runs. And, if only first books with substantial praise are examined, we know the second books will have a lower "batting average," since there is nowhere to go but down. There would be no way for all 24 authors to maintain their perfect record.

While the majority of second novels were not as praised, there are a handful of successes among this larger bunch. Marilynne Robinson's first novel, *Housekeeping*, was published in 1980 and was nominated for the Pulitzer Prize. Her next novel, which came out 24 years later and with about 25% more words, won the Pulitzer. This doesn't amount to enough data to say book inflation is bad. All that is certain is that even among literary authors, word counts creep upward after a first hit.

Whether you are J. K. Rowling or Amy Tan, book inflation is common. Stephen King's debut book, *Carrie*, was a massive hit and since then he has published over fifty novels under his own name. Only three have been shorter than *Carrie*. Romance writer Nicholas Sparks was an unknown when he was paid $1 million for his first published novel, *The Notebook*. His 17 books since have all been at least 25% longer. A successful author can publish books of whatever length and scope they please, and if their goal is to explore a bigger creative space, then perhaps the extra length is needed in many cases. But no matter who you are, before you write in that extra character or plot twist, it might be smart to take note of the trend that book inflation reveals—and perhaps to remember the simplicity that got you to great heights in the first place.

Chapter 9

Beginnings and Endings

*"That was the first sentence. The problem was
that I just couldn't think of the next one."*

—CHARLIE IN *THE PERKS OF BEING A
WALLFLOWER*, BY STEPHEN CHBOSKY

What makes a great opening sentence?

In response to a question on Twitter about her favorite first sentence in literature, novelist Margaret Atwood answered: "'Call me Ishmael.' Three words. Power-packed. Why Ishmael? It's not his real name. Who's he speaking to? Eh?"

Many consider the opening to *Moby Dick* to be the best first sentence of all time. It's one of a handful of openers that people are expected to know and recognize. Almost any list of either the best or most iconic novel first sentences ever written will include *Call me Ishmael.*

Atwood's justification—"Three words. Power-packed"—tapped into a common sensibility. Stephen King, in a 2013 interview with *The Atlantic*, cited his favorite three openers, and they averaged a mere six words long. Brevity can make for a phenomenal opening.

But what do the numbers say? In each book, a writer only gets one first sentence. What do you do with that opportunity? Do you go long or do you keep it short? Is one better, statistically, than the other?

As a first step, I looked at what each writer in my sample has chosen to do throughout his or her career. The results, to no surprise, vary enormously.

The median opener in Atwood's 15 novels is a compact nine words. Her first lines to *The Handmaid's Tale* ("We slept in what had once been the gymnasium") and *MaddAddam* ("In the beginning, you lived inside the Egg") are typical of her style. And openers like "Snowman wakes before dawn" and "I don't know how I should live" lower her tally even further.

Nine words, as a median length, is on the extreme low end of the scale when it comes to first sentences. Atwood's median is just one-third that of a writer like Salman Rushdie, whose median opener measures 29 words. Here's his 29-word opening to *Shame*.

> In the remote border town of Q, which when seen from the air resembles nothing so much as an ill-proportioned dumb-bell, there once lived three lovely, and loving, sisters.

That's a lot more clauses than any of Atwood's. Much like Rushdie, Michael Chabon goes long, holding a median length of 28 words. Out of the seven novels he has written, three have had openers of lengths of 41, 52, and 62 words. Only one of the seven has an opening sentence shorter than the average sentence length in the rest of the book.

Chabon and Rushdie are closer to typical than Atwood. Compared to the 31 other writers in my sample with at least five books to their name, Atwood comes in as the writer with the second-shortest first sentences, just behind Toni Morrison.

Some authors have no set pattern in their first sentences: The first sentence in Dickens's *A Tale of Two Cities* (the one that starts off "It was the best of times, it was the worst of times") measures in at 119 words, 17 commas, and an em-dash. The first sentence to his

AUTHORS WITH SHORTEST FIRST SENTENCES	MEDIAN LENGTH
Toni Morrison	5
Margaret Atwood	9
Mark Twain	11
Dave Eggers	11
Chuck Palahniuk	11.5
AUTHORS WITH LONGEST FIRST SENTENCES	**MEDIAN LENGTH**
Jane Austen	32
Vladimir Nabokov	29
Salman Rushdie	29
Michael Chabon	28
Edith Wharton	28

Only authors with five books were included. The variation is high when each book only has one data point.

A Christmas Carol is six words: "Marley was dead: to begin with." Both sentences are classics.

But the data reveals definite choices in other authors' careers. Including her 2015 novel, *God Help the Child*, Toni Morrison has written 11 books. Here are her opening sentences—she likes to keep it brief.

Toni Morrison's First Sentences	
The Bluest Eye (1970)	Here is the house.
Sula (1973)	In that place, where they tore the nightshade and blackberry patches from their roots to make room for the Medallion City Golf Course, there was once a neighborhood.
Song of Solomon (1977)	The North Carolina Mutual Life Insurance agent promised to fly from Mercy to the other side of Lake Superior at three o'clock.
Tar Baby (1981)	He believed he was safe.
Beloved (1987)	124 was spiteful.
Jazz (1992)	Sth, I know that woman.
Paradise (1997)	They shoot the white girl first.
Love (2003)	The women's legs are spread wide open, so I hum.
A Mercy (2008)	Don't be afraid.
Home (2012)	They rose up like men.
God Help the Child (2015)	It's not my fault.

While individual authors vary widely in their choices about first sentences, a larger trend emerges when we look at the 31 authors as a whole. Of all the books they've written, 69% of these books begin with first sentences that are *longer* than the average sentence throughout the rest of the book. That is, the standard choice for authors is to write *long* rather than *short* for their openers.

Does this disprove the theory that short is better, boiling it down to just a matter of personal choice? Or does this mean that most authors should cut down on their winding introductions?

To find out, I decided to go one step further and look at what are considered the "best" opening lines in literature. I've gone through eight different publications that ranked the "best" or "most memorable" first sentences. Twenty openers were on four of these lists, a consensus. Below are those twenty openers.

The 20 Best First Sentences in Literature	
BOOK/AUTHOR	**FIRST SENTENCE**
Pride and Prejudice / Jane Austen	It is a truth universally acknowledged, that a single man in possession of a good fortune, must be in want of a wife.
Moby Dick / Herman Melville	Call me Ishmael.
Lolita / Vladimir Nabokov	Lolita, light of my life, fire of my loins.
The Bell Jar / Sylvia Plath	It was a queer, sultry summer, the summer they electrocuted the Rosenbergs, and I didn't know what I was doing in New York.
Anna Karenina / Leo Tolstoy	Happy families are all alike; every unhappy family is unhappy in its own way.
1984 / George Orwell	It was a bright cold day in April, and the clocks were striking thirteen.
The Adventures of Huckleberry Finn / Mark Twain	You don't know about me without you have read a book by the name of *The Adventures of Tom Sawyer,* but that ain't no matter.
Murphy / Samuel Beckett	The sun shone, having no alternative, on the nothing new.
Ulysses / James Joyce	Stately, plump Buck Mulligan came from the stairhead, bearing a bowl of lather on which a mirror and a razor lay crossed.
The Stranger / Albert Camus	Mother died today.

BOOK/AUTHOR	FIRST SENTENCE
Slaughterhouse-Five / Kurt Vonnegut	All this happened, more or less.
A Tale of Two Cities / Charles Dickens	It was the best of times, it was the worst of times, it was the age of wisdom, it was the age of foolishness, it was the epoch of belief, it was the epoch of incredulity, it was the season of Light, it was the season of Darkness, it was the spring of hope, it was the winter of despair, we had everything before us, we had nothing before us, we were all going direct to Heaven, we were all going direct the other way—in short, the period was so far like the present period, that some of its noisiest authorities insisted on its being received, for good or for evil, in the superlative degree of comparison only.
Gravity's Rainbow / Thomas Pynchon	A screaming comes across the sky.
One Hundred Years of Solitude / Gabriel García Márquez	Many years later, as he faced the firing squad, Colonel Aureliano Buendía was to remember that distant afternoon when his father took him to discover ice.
The Trial / Franz Kafka	Someone must have slandered Josef K., for one morning, without having done anything truly wrong, he was arrested.
The Catcher in the Rye / J. D. Salinger	If you really want to hear about it, the first thing you'll probably want to know is where I was born, and what my lousy childhood was like, and how my parents were occupied and all before they had me, and all that David Copperfield kind of crap, but I don't feel like going into it, if you want to know the truth.
A Portrait of the Artist as a Young Man / James Joyce	Once upon a time and a very good time it was there was a moocow coming down along the road and this moocow that was coming down along the road met a nicens little boy named baby tuckoo.
Their Eyes Were Watching God / Zora Neale Hurston	Ships at a distance have every man's wish on board.
The Voyage of the Dawn Treader / C. S. Lewis	There was a boy called Eustace Clarence Scrubb, and he almost deserved it.
The Old Man and the Sea / Ernest Hemingway	He was an old man who fished alone in a skiff in the Gulf Stream and he had gone eighty-four days now without taking a fish.

Note: Not all the novels above were originally published in English. A translated version was used to compile the length comparison statistics.

Again, the variation is very high, with several ultra-short sentences and several winding ones, such as *The Catcher in the Rye*'s 63-word opener. The median length of these openings is

16 words. Although it's a painfully small sample, we do find that of the top 20 openers 12 (60%) were shorter than the average sentence in their respective books. In the broader sample used above, just 31% had an opener shorter than the book's average sentence.

It seems that Atwood and King were on to something when praising the power of a concise, direct opening: Shorter openings do tend to be more memorable.

But it's also important to note that the data in this section is saying far more than just *short is good*. The first sentence is only as popular as the rest of the book, and brevity alone will not make a first sentence great. In addition to "Call me Ishmael" Herman Melville also started novels off with the quick "Six months at sea!" and "We are off!" But nobody goes around praising the openings of *Typee* or *Mardi*.

Looking at the top twenty sentences, we can see that what each of the best openers have in common is not *length* but a certain originality or novelty that makes them memorable. This can be accomplished briefly or at length—but if anything it's the shock of the unpredicted and the unpredictable that makes them stick.

While looking at the stats can help us shed light on the question of what makes a good opener, it can't solve the question for us—because the great openers succeed by intentionally breaking patterns. Perhaps Stephen King put it best: In his *Atlantic* interview he said, "There are all sorts of theories and ideas about what constitutes a good opening line. . . . To get scientific about it is a little like trying to catch moonbeams in a jar."

It Was a Gloriously Sunny Day

We may not be able to reverse-engineer the perfect opener, but what do we find if we look at the other end of the spectrum—at *bad* first sentences? There are certain tropes, for instance, that have fallen into cliché, and that many an author would be wise to avoid.

At the very top of the list is: *It was a dark and stormy night* . . .

It's one of the most famous openings in literature, and it was original when it was first penned, almost 200 years ago by Edward Bulwer-Lytton in his 1830 novel *Paul Clifford*. Here's its complete form:

> It was a dark and stormy night; the rain fell in torrents, except at occasional intervals, when it was checked by a violent gust of wind which swept up the streets (for it is in London that our scene lies), rattling along the house-tops, and fiercely agitating the scanty flame of the lamps that struggled against the darkness.

But the sentence has since become an object of ridicule. Ray Bradbury mocked it in the opening to his novel *Let's All Kill Constance*:

> It was a dark and stormy night.
>
> Is that one way to catch your reader?
>
> Well, then, it was a stormy night with dark rain pouring in drenches on Venice. . . .

Likewise, for the last thirty years, the San Jose State University English department has been hosting a Bulwer-Lytton Fiction Contest, in which competitors try to create the worst opening sentence "to the worst of all possible novels."

Here's a recent winner:

> Folks say that if you listen real close at the height of the full moon, when the wind is blowin' off Nantucket Sound from nor' east and the dogs are howlin' for no earthly reason, you

can hear the awful screams of the crew of the "Ellie May," a sturdy whaler captained by John McTavish; for it was on just such a night when the rum was flowin' and, Davey Jones be damned, big John brought his men on deck for the first of several screaming contests.

Bulwer-Lytton got the short end of the stick by having the contest named after him. And while it might be unfair to call out an author from 1830 for falling into a cliché before it existed in full, it has become an old trope to open a book with a description of the weather. Elmore Leonard's first rule on writing, in fact—his *number one* rule—was "Never open a book with weather."

I wanted to find out who, among the authors we've looked at in this book, relied on the weather most often to set the scene. In other words, who is our Bulwer-Lytton award winner?

When I ran the numbers, a runaway champion leapt out from the pack. It turns out that when it comes to breezy weather openings, our leader also happens to be the author who has sold more books than any other living writer: Danielle Steel. (If you haven't read her, check out the line-up of romance novels at your local grocery store. She'll be there.)*

Here, in the first sentence of her first novel, she started to dabble with the weather opening:

It was a gloriously sunny day and the call from Carson Advertising came at nine-fifteen.

And it worked. So why not go back to the well? By 2014, Danielle Steel had written 92 novels, and a shocking 42 of these mention weather in the opening sentence. Read through them on the following pages at your own discretion.

---

* Book sales are often self-reported by publishers. A 2010 Canadian Broadcasting Company article stated 500 million copies of Steel's books have been sold.

It was a gloriously sunny day and the call from Carson Advertising came at nine-fifteen. | The weather was magnificent. | The early morning sun streamed across their backs as they unhooked their bicycles in front of Eliot House on the Harvard campus. | Hurrying up the steps of the brownstone on East Sixty-third Street, Samantha squinted her eyes against the fierce wind and driving rain, which was turning rapidly into sleet. | When it snows on Christmas Eve in New York, there is a kind of raucous silence, like bright colors mixed with snow. | The house at 2129 Wyoming Avenue, NW, stood in all its substantial splendor, its gray stone facade handsomely carved and richly ornate, embellished with a large gold crest and adorned with the French flag, billowing softly in a breeze that had come up just that afternoon. | The sun sank slowly onto the hills framing the lush green splendor of the Napa Valley. | The heat of the jungle was so oppressive that just standing in one place was almost like swimming through thick, dense air. | The sun reverberated off the buildings with the brilliance of a handful of diamonds cast against an iceberg, the shimmering white was blinding, as Sabina lay naked on a deck chair in the heat of the Los Angeles sun. | Everything in the house shone as the sun streamed in through the long French windows. | The rains were torrential northeast of Naples on the twenty-fourth of December 1943, and Sam Walker huddled in his foxhole with his rain gear pulled tightly around him. | Zoya closed her eyes again as the troika flew across the icy ground, the soft mist of snow leaving tiny damp kisses on her cheeks, and turning her eyelashes to lace as she listened to the horses' bells dancing in her ears like music. | The birds were already calling to each other in the early morning stillness of the Alexander Valley as the sun rose slowly over the hills, stretching golden fingers into a sky that within moments was almost purple. | The snowflakes fell in big white clusters, clinging together like a drawing in a fairy tale, just like in the books Sarah used to read to the children. | It was a chill gray day in Savannah, and there was a brisk breeze blowing in from the ocean. | The air was so still in the brilliant summer sun that you could hear the birds, and every sound for miles, as Sarah sat peacefully looking out her window. | The sky was a brilliant blue, and the day was hot and still as Diana Goode stepped out of the limousine with her father. | Charles Delauney limped only slightly as he walked slowly up the steps of Saint Patrick's Cathedral, as a bitter wind reached its icy fingers deep into his collar. | It was one of those perfect, deliciously warm Saturday afternoons in April, when the air on your cheek feels like silk, and you want to stay outdoors forever. | The weather in Paris was unusually warm as Peter Haskell's plane landed at Charles de Gaulle Airport. | The sounds of the organ music drifted up to the Wedgwood blue sky. | In the driving rain of a November day, the cab from London to Heathrow took forever. | The box arrived on a snowy afternoon two weeks before Christmas. | It was a brilliantly sunny day in New York, and the temperature had soared over the hundred mark long before noon. | The call came when she least expected it, on a snowy December afternoon, almost exactly thirty-four years after they met. | Marie-Ange Hawkins lay in the tall grass, beneath a huge, old tree, listening to the birds, and watching the puffy white clouds travel across the sky on a sunny August morning. | The sun glinted on the elegant mansard roof of The Cottage, as Abe Braunstein drove around the last bend in the seemingly endless driveway. | It was a perfect balmy May evening, just days after spring had hit the East Coast with irresistible appeal. | The sun was shining brightly on a hot June day in San Dimas, a somewhat distant suburb of L.A. | It was one of those chilly, foggy days that masquerade as summer in northern California, as the wind whipped across the long crescent of beach, and whiskbroomed a cloud of fine sand into the air. | It was a lazy summer afternoon as Beata Wittgenstein strolled along the shores of Lake Geneva with her parents. | The sailing yacht *Victory* made her way elegantly along the coast toward the old port in Antibes on a rainy November day. | The sun was brilliant and hot, shining down on the deck of the motor yacht *Blue Moon*. | Olympia Crawford Rubinstein was whizzing around her kitchen on a sunny May morning, in the brownstone she shared with her family on Jane Street in New York, near the old meat-packing district of the West Village. | It was a beautiful hot July day in Marin County, just across the Golden Gate Bridge from San Francisco, as Tanya Harris bustled around her kitchen, organizing her life. | It was a quiet, sunny November morning, as Carole Barber looked up from her computer and stared out into the garden of her Bel-Air home. | It was an absolutely perfect June day as the sun came up over the city, and Coco Barrington watched it from her Bolinas deck. | Hope Dunne made her way through the silently falling snow on Prince Street in SoHo in New York. | Seth Adams left Annie Ferguson's West Village apartment on a sunny September Sunday afternoon. | There was a heavy snowfall that had started the night before as Brigitte Nicholson sat at her desk in the admissions office of Boston University, meticulously going over applications. | The two men who lay parched in the blistering sun of the desert were so still they barely seemed alive. | Lily Thomas lay in bed when the alarm went off on a snowy January morning in Squaw Valley.

Compared to all other authors I searched through, including all 26 authors in my sample who have penned at least ten books, no one comes close to Steel's weather rate. Even compared to her closest peers who rival her popularity and prolific nature, Steel is far and away the leader of using weather to open her books.

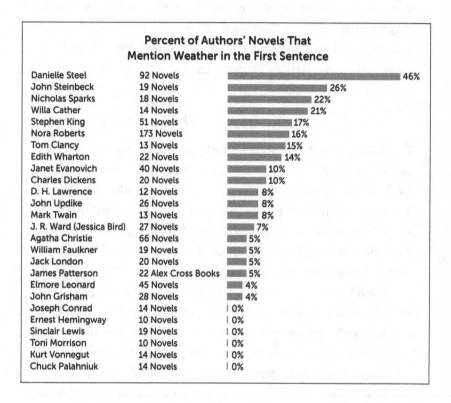

Percent of Authors' Novels That Mention Weather in the First Sentence

Author	Novels	Percent
Danielle Steel	92 Novels	46%
John Steinbeck	19 Novels	26%
Nicholas Sparks	18 Novels	22%
Willa Cather	14 Novels	21%
Stephen King	51 Novels	17%
Nora Roberts	173 Novels	16%
Tom Clancy	13 Novels	15%
Edith Wharton	22 Novels	14%
Janet Evanovich	40 Novels	10%
Charles Dickens	20 Novels	10%
D. H. Lawrence	12 Novels	8%
John Updike	26 Novels	8%
Mark Twain	13 Novels	8%
J. R. Ward (Jessica Bird)	27 Novels	7%
Agatha Christie	66 Novels	5%
William Faulkner	19 Novels	5%
Jack London	20 Novels	5%
James Patterson	22 Alex Cross Books	5%
Elmore Leonard	45 Novels	4%
John Grisham	28 Novels	4%
Joseph Conrad	14 Novels	0%
Ernest Hemingway	10 Novels	0%
Sinclair Lewis	19 Novels	0%
Toni Morrison	10 Novels	0%
Kurt Vonnegut	14 Novels	0%
Chuck Palahniuk	14 Novels	0%

Steel uses the weather trope almost twice as often as her closest competitor. By my count, half the time the weather is dreary, rainy, or stormy in some way. The other half, the weather is of the more pleasant or perfect variety ("perfect deliciously warm Saturday afternoons," "perfect balmy May evening," "absolutely perfect June day"). It's still working for her, even if it verges on some Bulwer-Lytton Award–winning candidates (my favorite is simply: "The weather was magnificent").

As much grief as the old weather opening gets, it remains a standby for many authors. Sometimes you just have to set the scene, after all. Consider the opening to *1984*, considered one of the greatest first sentences of all time: "It was a bright cold day in April, and the clocks were striking thirteen." Weather isn't necessarily a death-knell, especially when (as Orwell does) it's used to play with the reader's expectations. Even in the most well-regarded writing, weather is still a common opening motif: In the 86 Pulitzer Prize winners for fiction, 13 openings rely on the weather.

The Cliffhanger Boys

Franklin Dixon got his start in the book-publishing world in 1927 as the author of *The Tower Treasure*, the first book in the Hardy Boys series. Dixon found immediate success with *The Tower Treasure* and its sequels. The first five books in the Hardy Boys series all ranked in the top 200 on *Publishers Weekly* 2001 list of "All-Time Bestselling Children's Books." Dixon went on to write more than 300 more Hardy Boys books, up until *Movie Mission* in 2011.

But if Franklin Dixon was old enough to write the first Hardy Boys novel in 1927, how was he still writing books 84 years later?

The truth is there was no person named Franklin Dixon. The name was a creation of the Stratemeyer Syndicate, a book-packaging company created by Edward Stratemeyer. In its first fifty years of operation following its creation in 1899, the company cranked out 98 different series. Everything about these books was thought out in detail, including the use of pseudonyms like Franklin Dixon. By writing under a single fake name, the ghostwriters had little control over the series or their payments. If a writer wished to discontinue writing, the series could continue without anyone (mostly the children readers) having any idea that there had been a change.

We've already seen that the series was able to hold together for so long in part because of strict rules about length and structure. But these were far from the only rules. Ghostwriters were given plot outlines by the syndicate (often penned by Edward Stratemeyer) and told to keep their writing within certain guidelines.

One of the most important rules was that each chapter had to end mid-action—with a cliffhanger.

Consider the first seven Hardy Boys books. If you pick up any of these books today you'll notice they all have exactly 20 chapters.* They are all between 32,000 and 36,000 words. And they all like to end their chapters mid-action. Below is a chart showing the last sentence in each chapter of the first Hardy Boys book, *The Tower Treasure*.

CHAPTER	LAST SENTENCE
1	In a few moments the boys were tearing down the road in pursuit of the automobile!
2	The same thought was running through Frank's and Joe's minds: maybe this mystery would turn out to be their first case!
3	"Follow me!"
4	The detective stood by sullenly as Frank pulled out a penknife and began to scrape the red paint off part of the fender.
5	"Come on, everybody!"
6	"My dad is innocent!"
7	"Slim, we'll do all we can to help your father."
8	"And this time a swell one!"
9	And were they about to share another of his secrets?
10	"But where is he now?"
11	"Mother's pretty worried that something has happened to Dad."
12	With that he arose, stumped out of the room, and left the house.
13	"Where's some water?"
14	The brothers raced from the house, confident that they were about to solve the Tower Treasure mystery.
15	Frank and Joe, tingling with excitement, followed.
16	"I've found a buried chest!"
17	It came from the top room of the old tower!
18	"This must be the place!"
19	Without warning the trap door was slammed shut and locked from the outside!
20	"An excellent idea!"

---

* The Hardy Boys books have been revised since their original publication. This chapter used the revised versions. The originals were standardized as well at 25 chapters each.

Skimming the list, it's easy to see some patterns, especially when it comes to punctuation. Fourteen of twenty chapters end with either an exclamation point or a question mark. The cliffhanger rule is maintained in a very obvious way: either through obvious excitement (!) or obvious mystery (?).

The Hardy Boys are not exactly a model of subtle writing. Earlier in this book we looked at Elmore Leonard's advice about exclamation points: "no more than two or three per 100,000 words." The Hardy Boys use more than 900 exclamation points per 100,000 words.

But even if you compare a standard sentence in the Hardy Boys to one of the chapter closers, it's clear that the chapters are built to go out with a bang. To simplify things let's call any of the following an "Obvious Cliffhanger Mark": an exclamation point, a question mark, an em-dash (as in a quote cutoff mid-sentence), or an ellipsis. In the Hardy Boys about 19% of all sentences end with one of these "Obvious Cliffhanger Marks." But, among sentences that close out a chapter, 71% end with the "Obvious Cliffhanger Mark."

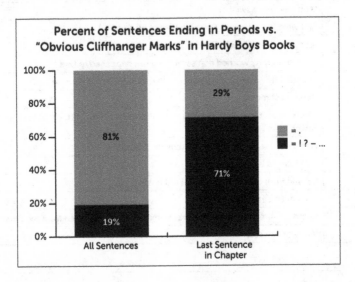

The Nancy Drew books, also produced by the Stratemeyer Syndicate, follow the same level of consistency. Looking at the first seven books as a sample, they are all twenty chapters long and all have between 32,000 and 37,000 words, and they use cliffhangers in the same way.

Below is the same plot for the Nancy Drew books. Chapter endings are almost four times as likely to end with an "Obvious Cliffhanger Mark" compared with the average sentence.

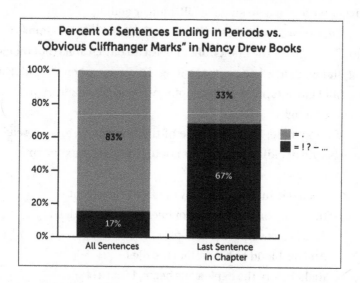

Books run by the Stratemeyer Syndicate weren't the only ones who used this tactic to make children keep reading. Enid Blyton is a prolific British children's author who wrote 186 novels, starting in 1922, that would end up selling half a billion copies. The Famous Five, a Blyton series about five kids who go on adventures when back from boarding school, had chapters that ended with the "Obvious Cliffhanger Mark" 83% of the time. A normal sentence in her book only ended in such excitement 25%.

In today's popular children's and young adult literature the trend is all but absent. Harry Potter ends with such obvious cliffhangers about 14% of the time. The Goosebumps series clocks in at 18%, the Hunger Games series at 4%, and all 142 chapters of the Divergent trilogy end in a period.

It's impossible to come up with an objective measure of cliff-hanger-ness, though. The Hunger Games may not use loud punctuation at the end of the chapters, but that doesn't mean there are not cliffhangers. Ending chapter after chapter with a question mark or exclamation point may be too heavy-handed for modern sensibilities. But there is another style choice that almost all popular page-turner writers use to signal a cliffhanger ending.

There are 82 chapters in Suzanne Collins's Hunger Games trilogy. Throughout the books, about 9% of all paragraphs (not including dialogue) are one sentence long. However, when considering the final paragraph of each chapter, we find that 62% are just one sentence long.

For example, here are some of the last *paragraphs* a reader sees before deciding whether or not to flip to the next chapter.

> Then the ants bore into my eyes and I black out.
> In other words, I step out of line and we're all dead.
> This is one of his death traps.
> And his blood as it splatters the tiles.
> Right before the explosions begin, I find a star.

It's hard to get the full effect of such a short chapter ender from the single sentences above. The average paragraph in the Hunger Games is ninety words. It fills more than one-third of a typical page. But for the ends of her chapters, Collins chooses to avoid anything that might look like a wall of text. She gives the reader a short, attention-grabbing plot point to keep them interested.

Collins's practice of ending chapters on a short, punchy paragraphs turns out to be near universal among thriller writers. Patterson's 22 Alex Cross books all have shorter chapter-ending paragraphs than the paragraphs throughout. Stephen King is twice as likely to use a single-sentence paragraph when it's the last paragraph of a chapter.

Abrupt Chapter Endings of Popular Thriller, Suspense, and Mystery Writers			
AUTHOR	BOOKS	PERCENT OF ONE-SENTENCE CLOSING PARAGRAPHS	OVERALL PERCENT OF ONE-SENTENCE PARAGRAPHS
Suzanne Collins	3 Hunger Games	62%	9%
Dan Brown	4 Robert Langdon Books	53%	39%
James Patterson	22 Alex Cross Books	57%	26%
Clive Cussler	23 Dirk Pitt Novels	48%	23%
David Baldacci	29 Novels	56%	37%
Stephen King	51 Novels	50%	26%
Gillian Flynn	3 Novels	40%	27%
Michael Crichton	24 Novels	54%	33%
Tom Clancy	13 Novels	19%	11%
Veronica Roth	3 Divergent Books	52%	25%

Note: Some books have clearly titled chapters. Others have page breaks with no markings and some have no page breaks at all. For some books, judgment calls were required to decide what constituted a chapter.

Of the last 40 *New York Times* number one bestsellers that were thrillers, mystery, or suspense, 36 books had chapter-ending paragraphs that were shorter than the paragraphs within the rest of the chapter. The typical thriller has 60% more one-sentence paragraphs at the end of the chapter than in the middle.

Not everyone is a fan of these abrupt paragraphs. In a review of Michael Crichton's *Jurassic Park* Martin Amis criticized the novel for having "one-page chapters, one-sentence paragraphs and one-word sentences." Even if Amis was not talking about the last chunk of text in a chapter, it's clear he considers Crichton's use of short paragraphs a cliché of the thriller genre.

But in *The Information*, Amis's novel that came out the same year as his *Jurassic Park* review, he ends 30% of his chapters with one-sentence paragraphs. That's almost twice as often as he uses a one-sentence prose paragraph in the rest of the book. Even some literary writers like Amis find a certain usefulness in ending abruptly to get the reader to keep reading.

The page-turner practice has not in general crept its way from thrillers to literary fiction. In 2013 and 2014, 41 different novels were given at least one of the following honors: *New York Times* Top Ten Books of the Year, Pulitzer Prize finalists, Man Booker Prize short list, National Book Award finalists, National Book Critics Circle finalists, and *Time* magazine's best books of the year. Thirty-eight of these books had chapters. Of these books, just twenty had more one-sentence paragraphs at the end of chapters than in the middle of the chapter. That's close to half, about what could be considered random.

While there's no sign that literary writers en masse will be using the device anytime soon, among popular thriller writers the one-sentence ender seems to be the natural evolution of the Hardy Boys' or Enid Blyton's Famous Five's distinctive cliffhanger marks. There's one very good reason why page-turners keep seeking out punchy endings, and it's the same reason that the Hunger Games and Alex Cross series each have sold millions of copies. What is it?

Cliffhangers work.

Epilogue

Throughout this book, I've been looking for the blend of rules and rule-breaking that come together to make writing work—or rather to make writing excel. It's an odd mix of consistency and the unexpected, of simple communication and whimsical delight, but it's what we find driving our best fiction forward.

I grew up on Roald Dahl books. *Charlie and the Chocolate Factory*, *The BFG*, *Matilda*, and *The Twits* were some of the most popular books in my elementary school despite the fact that we were reading them forty years after they were penned. But as I was in the midst of researching and writing this book there was one forgotten Dahl short story that I found myself rereading over and over: "The Great Automatic Grammatizator."

An engineer named Adolph Knipe dreams of writing stories that people will read. Knipe looks at the beginning of his latest failed novel attempt, which begins, of course, "The night was dark and stormy." Then he has a eureka moment.

> . . . he was struck by a powerful but simple little truth, and it
> was this: **That English grammar is governed by rules that**

are almost mathematical in their strictness! Given the words, and given the sense of what is to be said, then there is only one correct order in which those words can be arranged.

He invents a machine he calls the Great Automatic Grammatizator. It takes in plots and can spit out a finished story. Knipe starts with short stories. Soon he dials up the complexity to the length of the novel and, getting more daring, he programs it to write a "high class intelligent book." Knipe harnesses his machine to the point where "one half of all novels" published in English are written by the Great Automatic Grammatizator.

Now a tycoon due to the power of his bestseller machine, he forces other writers to the brink of starvation. The narrator of this story is not the engineer Knipe but an opposing author. Knipe offers the narrator a contract *not* to write. This would allow the narrator to eat, but the automated computer-generated stories would take over. Or, the narrator can decide not to sign the contract, which would allow him to write but leave him broke. The story ends with a plea from the narrator: "Give us strength, Oh Lord, to let our children starve."

Like "The Great Automatic Grammatizator," my book has been about the marriage of numbers and words. People often have polarizing reactions when objective analysis is applied to art. As I've discussed this book, I've encountered two opposite camps, which I've categorized in my head as the extreme skeptic and the doomsdayer.

The extreme skeptic is uneasy every time they see a number next to a word. Writing is an art, not a science, so how can math provide any substance?

If you've made it this far in the book I hope I have convinced you not to be that extreme skeptic. I've tried to tackle questions that are common to readers and writers. There's a distinct benefit to being able to run through millions of words at once. You may lose a word's impact on a particular page, but a new appreciation for an author can come to light. Patterns that are spread out over a corpus of literature, too large to be consumed by any one reader, can teach new trends, ideas, techniques, and wisdom that would otherwise be hidden.

In contrast to the skeptic, there is the person yelling that the sky is falling whenever they see anything to do with numbers and texts. If numbers can help us predict what will be popular to read, when will an algorithm just start writing novels for us? This is "The Great Automatic Grammatizator" distilled.

Even today, more than sixty years after "The Great Automatic Grammatizator" was published, the concept is far-fetched science fiction. The numbers I've looked at here, and the calculations I've used, can help us read and see patterns—but they can't help us know when to break them. The questions I sought to answer in my book are primitive. Are there words worth avoiding? How do popular authors use certain words? What are the most substantial differences in the way people from different backgrounds write?

These questions are only a starting point for writers or readers—not an attempt to "engineer" art as much as a way to understand it or describe it. If you were an aspiring painter in 1900 you might want to know the specific paints and techniques that Monet was starting to use. If you were a band in the 1960s you would want to know how the Beatles were recording their songs. In either case, you would want to understand the craft in detail and on a technical level before making your own masterpiece. Reading a book is the easiest way to understand how a novel is crafted. Examining the patterns of thousands of books is going to answer different questions, but it is likewise a useful way to understand how books are truly crafted.

Somewhere between the skeptic and doomsdayer is where I hope you have landed after reading this book. Successful writers pen hundreds of thousands of words in their lifetime. In any other field with hundreds of thousands of data points it would be quite clear that the information could be mined to examine human behavior and psychology. I believe the same is true for examining words.

When Frederick Mosteller and David Wallace used equations to determine the authorship of *The Federalist Papers*, they were solving one small question about writing. It was a question with a definite answer, which made it simpler, but it showed that information that may not be obvious on first read is right there, hiding in plain sight.

The written word and the world of numbers should not be kept apart. It's possible to be a lover of both. Through the union of writing and math there is so much to learn about the books we love and the writers we admire. And by looking at the patterns, we can appreciate that beautiful moment where the pattern breaks, and where a brilliant new idea bursts into the world.

Acknowledgments

I am fortunate that Simon & Schuster gave me the chance to write this book, and I am fortunate for so many people who helped me during this long process.

I want to thank my agent, Jackie Ko, who invested invaluable time and energy into all my creative pursuits from the beginning.

I also want to thank my editor, Jonathan Cox, who made this project happen. I am beyond grateful for his input, which enhanced the book at every level.

Eric Brewster was the first person to lay eyes on any of this material. I want to thank him for letting me know which of my stream-of-consciousness ideas should stay in my head and which should make it to the page.

I want to thank my parents, Stephen Blatt and Faith Minard, for instilling in me the virtues needed to create this book. I'd also like to thank my older brother, Zach Blatt, who has influenced me since I was born.

Tony Khan was beyond helpful, and I thank him for being a great friend.

I want to thank everyone, living and dead, from the *Harvard*

Lampoon. Without being a part of such a sharp community of writers, I never would have developed the mindset or skills required to complete this book.

In addition, I want to thank all these amazing people, both for their generous input and for their hospitality. Thanks to Andy Spielvogel, Zack and Jody Wortman, Kurt Slawitschka, Katie Ryan, Ben Silva, Eric Arzoian, Florian Mayr, Daniel Claridge, Peter Manges, Sierra Katow, Daniel Bredar, Tyler Richard, Ethan Glasserman, E. J. Bensing, Meryl Natow, Ari Rubin, James Yoder, Jeffrey Hajdin, Nicole Levin, and J. J. Shpall.

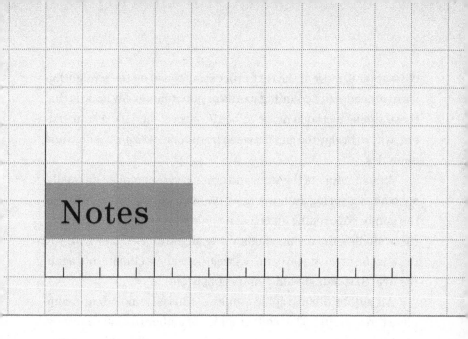

Notes

This book has relied on a variety of samples and methods, the essentials of which I've tried to include within each chapter along the way. But several examples warrant further detail, as do the larger book lists themselves and the nature of the decisions I made throughout the writing process.

In total, approximately 1,500 books (not including tens of thousands of full-length fan fiction or Literotica novels) were collected and converted into raw text files for this project. Dedications, copyrights, prefaces, acknowledgments, and so on at the beginning and end of the books were edited out of the files, but text within the books was not touched, even if not strictly part of the author's prose (for example, a head like "Chapter 10: Mayhem at the Ministry" would remain). Running the same analysis on different editions of the same book could, for this reason, produce slightly different results. To combat this, I focused only on trends or patterns where the statistical difference was significant enough that book edition would have no meaningful impact on the results.

Most of the text processing was completed in Python and with the help of NLTK (Natural Language Toolkit). With the exception of Chapter 3, most of the text processing relied on the straightforward procedure of counting words or punctuation. My hope is that these simple methods prove to be the easiest, and most transparent, way to bridge the gap between traditional writing wisdom and analytics.

A fair amount of the stats and graphs were compiled manually, such as measuring author names on book covers or determining the number of opening lines that include weather. If you duplicated these studies, the results might differ slightly based on where you draw a box or what you count as weather versus climate, but again the overall trends should remain consistent.

All author bibliographies, unless otherwise noted, are complete through 2014. The bibliographies of authors include only novels, barring a few rare exceptions. I chose to omit authors' writing in other genres (nonfiction, memoir, short stories, etc.) so as to give the main sample a central focus and consistency. Also, on a logistical level, while it was possible to collect all novels notable authors have written, some have bibliographies of short stories so large it would be difficult to be certain that all stories were included.

There are a few exceptions, mentioned in the text, in which the set of texts used for an author is not their full novel bibliography. For some authors, like J. K. Rowling, I ignored smaller works if they are known mostly for one series. And for a handful of authors who are known primarily for their more stylized nonfiction, like Mitch Albom or Truman Capote, I also included their narrative works of nonfiction.

I acknowledge that there were subjective judgments in assembling these bibliographies. There is no set definition of novel versus novella or a clean line between fiction and narrative nonfiction. Some books, for example Ray Bradbury's *The Martian Chronicles*, are structured in such a way that some would consider them short story collections, while others consider them episodic novels. Authors who write books with co-authors only make things more difficult. Many authors have had books published after their death. In some cases, these were finished books that happened to be released right after passing. In other cases, they are half-finished works.

I often made judgment calls on whether or not to include a book in the bibliography, which you can explore in the following pages. An author might appear slightly up or down on a given chart if one or two books had been removed from or added to their bibliography. But I concentrated on larger trends in the text, and these larger trends hold even with slight variations to the original samples.

Data Sources

Immediately following are the bibliographies for all the authors used in this book. Any statement in this book that reads "Jane Austen used the word X at rate Y" means "Of the six Jane Austen books listed in this notes section, she used the word X at rate Y."

Following the author bibliographies I have also listed specialized samples for which I explained the methodology throughout the main text, but not the exact contents. For example, "Modern Literary Fiction" was categorized as books that won various prizes over a set time period, but the text did not have room to include the full listing. Each of those titles can be found in this Notes section.

Chinua Achebe—5 Novels

Things Fall Apart
No Longer at Ease
Arrow of God
A Man of the People
Anthills of the Savannah

Douglas Adams—7 Novels

The Hitchhiker's Guide to the Galaxy
The Restaurant at the End of the Universe
So Long, and Thanks for All the Fish
Dirk Gently's Holistic Detective Agency
The Long Dark Tea-Time of the Soul
Mostly Harmless
Life, the Universe and Everything

Mitch Albom—6 Books

Tuesdays with Morrie

The Five People You Meet in
   Heaven

For One More Day

Have a Little Faith

The Time Keeper

The First Phone Call from Heaven

Isaac Asimov—7 Foundation Series Novels

Prelude to Foundation

Forward the Foundation

Foundation

Foundation and Empire

Second Foundation

Foundation's Edge

Foundation and Earth

Jean Auel—6 Earth's Children Novels

The Clan of the Cave Bear

The Valley of Horses

The Mammoth Hunters

The Plains of Passage

The Shelters of Stone

The Land of Painted Caves

Jane Austen—6 Novels

Sense and Sensibility

Pride and Prejudice

Mansfield Park

Emma

Northanger Abbey

Persuasion

David Baldacci—29 Novels

The Camel Club

The Collectors

Stone Cold

Divine Justice

Hell's Corner

Split Second

Hour Game

Simple Genius

First Family

The Sixth Man

King and Maxwell

The Whole Truth

Deliver Us from Evil

Zero Day

The Forgotten

The Escape

The Innocent

The Hit

The Target

Absolute Power

Total Control

The Winner

The Simple Truth

Saving Faith

Wish You Well

Last Man Standing

The Christmas Train

True Blue

One Summer

Enid Blyton—21 Famous Five Novels

Five on a Treasure Island

Five Go Adventuring Again

Five Run Away Together

Five Go to Smuggler's Top

Five Go Off in a Caravan

Five on Kirrin Island Again

Five Go Off to Camp

Five Get into Trouble

Five Fall into Adventure

Five on a Hike Together

Five Have a Wonderful Time

Five Go Down to the Sea

Five Go to Mystery Moor

Five Have Plenty of Fun

Five on a Secret Trail

Five Go to Billycock Hill

Five Get into a Fix

Five on Finniston Farm

Five Go to Demon's Rocks

Five Have a Mystery to Solve

Five Are Together Again

Ray Bradbury—11 Novels

The Martian Chronicles

Fahrenheit 451

Dandelion Wine

*Something Wicked This Way
   Comes*

The Halloween Tree

Death Is a Lonely Business

A Graveyard for Lunatics

Green Shadows, White Whale

From the Dust Returned

Let's All Kill Constance

Farewell Summer

Ann Brashares—9 Novels

*The Sisterhood of the Traveling
   Pants*

*The Second Summer of the
   Sisterhood*

Girls in Pants

Forever in Blue

The Last Summer (of You and Me)

3 Willows

My Name Is Memory

Sisterhood Everlasting

The Here and Now

Charlotte Brontë—4 Novels

Jane Eyre

Shirley

Villette

The Professor

Dan Brown—4 Robert Langdon Novels

Angels & Demons

The Da Vinci Code

The Lost Symbol

Inferno

Truman Capote—5 Books

The Grass Harp

Over Voices, Other Rooms

Summer Crossing

Breakfast at Tiffany's

In Cold Blood

Willa Cather—12 Novels

Sapphira and the Slave Girl

One of Ours

My Ántonia

Shadows on the Rock

A Lost Lady

Lucy Gayheart

O Pioneers!

Death Comes for the Archbishop

The Song of the Lark

The Professor's House

My Mortal Enemy

Alexander's Bridge

Michael Chabon—7 Novels

The Mysteries of Pittsburgh

Wonder Boys

The Amazing Adventures of
    Kavalier & Clay

The Final Solution

The Yiddish Policemen's Union

Gentlemen of the Road

Telegraph Avenue

John Cheever—5 Novels

The Wapshot Chronicle

The Wapshot Scandal

Bullet Park

Falconer

Oh What a Paradise It Seems

Agatha Christie—66 Novels

The Mysterious Affair at Styles

The Secret Adversary

The Murder on the Links

The Man in the Brown Suit

The Secret of Chimneys

The Murder of Roger Ackroyd

The Big Four

The Mystery of the Blue Train

The Seven Dials Mystery

The Murder at the Vicarage

The Sittaford Mystery

Peril at End House

Lord Edgware Dies

Murder on the Orient Express

Why Didn't They Ask Evans?

Three Act Tragedy

Death in the Clouds

The A.B.C. Murders

Murder in Mesopotamia

Cards on the Table

Dumb Witness

Death on the Nile

Appointment with Death

Hercule Poirot's Christmas

Murder Is Easy

And Then There Were None

Sad Cypress

One, Two, Buckle My Shoe

Evil Under the Sun

N or M?

The Body in the Library

Five Little Pigs

The Moving Finger

Towards Zero

Death Comes as the End

Sparkling Cyanide

The Hollow

Taken at the Flood

Crooked House

A Murder Is Announced

They Came to Baghdad

Mrs McGinty's Dead

They Do It with Mirrors

After the Funeral

A Pocket Full of Rye

Destination Unknown

Hickory Dickory Dock

Dead Man's Folly

4:50 From Paddington

Ordeal by Innocence

Cat Among the Pigeons

The Pale Horse

The Mirror Crack'd from Side to Side

The Clocks

A Caribbean Mystery

At Bertram's Hotel

Third Girl

Endless Night

By the Pricking of My Thumbs

Hallowe'en Party

Passenger to Frankfurt

Nemesis

Elephants Can Remember

Postern of Fate

Curtain

Sleeping Murder

Tom Clancy—13 Novels

The Hunt for Red October
Patriot Games
The Cardinal of the Kremlin
Clear and Present Danger
The Sum of All Fears
Without Remorse
Debt of Honor

Executive Orders
The Bear and the Dragon
Red Rabbit
The Teeth of the Tiger
Red Storm Rising
Rainbow Six

Cassandra Clare—9 Novels

City of Bones
City of Ashes
City of Glass
City of Fallen Angels
City of Lost Souls

City of Heavenly Fire
Clockwork Angel
Clockwork Prince
Clockwork Princess

Suzanne Collins—3 Hunger Games Novels

The Hunger Games
Catching Fire

Mockingjay

Michael Connelly—27 Novels

The Black Echo
The Black Ice
The Concrete Blonde
The Last Coyote
The Poet
Trunk Music
Blood Work
Angels Flight
Void Moon
A Darkness More Than Night
City of Bones
Chasing the Dime
Lost Light
The Narrows

The Closers
The Lincoln Lawyer
Echo Park
The Overlook
The Brass Verdict
The Scarecrow
Nine Dragons
The Reversal
The Fifth Witness
The Drop
The Black Box
The Gods of Guilt
The Burning Room

Joseph Conrad—14 Novels

Nostromo

An Outcast of the Islands

The Rescue

Heart of Darkness

The Nigger of the "Narcissus"

Almayer's Folly

Lord Jim

Typhoon

Victory

The Arrow of Gold

Under Western Eyes

Chance

The Secret Agent

The Shadow-Line

Michael Crichton—24 Novels

Odds On

Scratch One

Easy Go

A Case of Need

Zero Cool

The Andromeda Strain

The Venom Business

Drug of Choice

Grave Descend

Binary

The Terminal Man

The Great Train Robbery

Eaters of the Dead

Congo

Sphere

Jurassic Park

Rising Sun

Disclosure

The Lost World

Airframe

Timeline

Prey

State of Fear

Next

Clive Cussler—23 Dirk Pitt Novels

Pacific Vortex

The Mediterranean Caper

Iceberg

Raise the Titanic!

Vixen

Night Probe!

Deep Six

Cyclops

Treasure

Dragon

Sahara

Inca Gold

Shock Wave

Flood Tide

Atlantis Found

Valhalla Rising

Trojan Odyssey

Black Wind

Treasure of Khan

Arctic Drift

Crescent Dawn

Poseidon's Arrow

Havana Storm

James Dashner—3 Maze Runner Novels

The Maze Runner
The Scorch Trials
The Death Cure

Don DeLillo—15 Novels

Americana
End Zone
Great Jones Street
Ratner's Star
Players
Running Dog
The Names
White Noise
Libra
Mao II
Underworld
The Body Artist
Cosmopolis
Falling Man
Point Omega

Charles Dickens—20 Novels

The Posthumous Papers of the Pickwick Club
The Adventures of Oliver Twist
The Life and Adventures of Nicholas Nickleby
The Old Curiosity Shop
Barnaby Rudge
A Christmas Carol
The Chimes
The Cricket on the Hearth
The Battle of Life
The Mystery of Edwin Drood
The Haunted Man and the Ghost's Bargain
The Life and Adventures of Martin Chuzzlewit
Dombey and Son
David Copperfield
Bleak House
Hard Times
Little Dorrit
A Tale of Two Cities
Great Expectations
Our Mutual Friend

Theodore Dreiser—8 Novels

Sister Carrie
Jennie Gerhardt
The "Genius"
The Financier
An American Tragedy
The Stoic
The Titan
The Bulwark

Jennifer Egan—4 Novels

The Invisible Circus
Look at Me
The Keep
A Visit from the Goon Squad

Dave Eggers—6 Novels

You Shall Know Our Velocity
What Is the What
The Wild Things
A Hologram for the King

The Circle
Your Fathers, Where Are They?
And the Prophets, Do They Live
Forever?

Jeffrey Eugenides—3 Novels

The Virgin Suicides
Middlesex

The Marriage Plot

Janet Evanovich—40 Novels

One for the Money
Two for the Dough
Three to Get Deadly
Four to Score
High Five
Hot Six
Seven Up
Hard Eight
To the Nines
Ten Big Ones
Eleven on Top
Twelve Sharp
Lean Mean Thirteen
Fearless Fourteen
Finger Lickin' Fifteen
Sizzling Sixteen
Smokin' Seventeen
Explosive Eighteen
Notorious Nineteen
Takedown Twenty
Top Secret Twenty-One

Visions of Sugar Plums
Plum Lovin'
Plum Lucky
Plum Spooky
Wicked Appetite
Wicked Business
Metro Girl
Motor Mouth
Back to the Bedroom
Smitten
Wife for Hire
The Rocky Road to Romance
Hero at Large
Foul Play
The Grand Finale
Thanksgiving
Manhunt
Love Overboard
Naughty Neighbor

William Faulkner—19 Novels

As I Lay Dying
The Sound and the Fury
Unvanquished
Sanctuary
Light in August
The Reivers
Go Down, Moses
Absalom, Absalom!
The Wild Palms
The Hamlet

The Mansion
Requiem for a Nun
Intruder in the Dust
Pylon
A Fable
The Town
Mosquitoes
Flags in the Dust
Soldiers' Pay

Joshua Ferris—3 Novels

Then We Came to the End
The Unnamed

To Rise Again at a Decent Hour

F. Scott Fitzgerald—4 Novels

The Great Gatsby
Tender Is the Night

This Side of Paradise
The Beautiful and Damned

Ian Fleming—12 James Bond Novels

Casino Royale
Live and Let Die
Moonraker
Diamonds Are Forever
From Russia, with Love
Dr. No

Goldfinger
Thunderball
The Spy Who Loved Me
On Her Majesty's Secret Service
You Only Live Twice
The Man with the Golden Gun

Gillian Flynn—3 Novels

Dark Places
Gone Girl

Sharp Objects

E. M. Forster—6 Novels

Maurice

A Passage to India

The Longest Journey

Howards End

A Room with a View

Where Angels Fear to Tread

Jonathan Franzen—4 Novels

The Twenty-Seventh City

Strong Motion

The Corrections

Freedom

Charles Frazier—3 Novels

Cold Mountain

Thirteen Moons

Nightwoods

William Gaddis—5 Novels

Agape Agape

The Recognitions

Carpenter's Gothic

A Frolic of His Own

JR

Neil Gaiman—7 Novels

Neverwhere

Stardust

American Gods

Coraline

Anansi Boys

The Graveyard Book

The Ocean at the End of the Lane

Mark Greaney—6 Novels

Support and Defend

Ballistic

On Target

The Gray Man

Full Force and Effect

Dead Eye

John Green—4 Novels

Looking for Alaska

An Abundance of Katherines

Paper Towns

The Fault in Our Stars

John Grisham—28 Novels

The Firm

The Pelican Brief

The Client

The Chamber

The Rainmaker

The Runaway Jury

The Partner

The Street Lawyer

The Testament

The Brethren

A Painted House

Skipping Christmas

The Summons

The Kings of Torts

Bleachers

The Last Juror

The Broker

Playing for Pizza

The Appeal

The Associate

Ford County

The Confession

The Litigators

Calico Joe

The Racketeer

Gray Mountain

A Time to Kill

Sycamore Row

Dashiell Hammett—5 Novels

Red Harvest

The Dain Curse

The Maltese Falcon

The Glass Key

The Thin Man

Nathaniel Hawthorne—6 Novels

Fanshawe

The Scarlet Letter

The House of the Seven Gables

The Blithedale Romance

The Marble Faun

Septimus Felton

Ernest Hemingway—10 Novels

To Have and Have Not

The Sun Also Rises

A Farewell to Arms

The Torrents of Spring

For Whom the Bell Tolls

Islands in the Stream

The Old Man and the Sea

The Garden of Eden

Across the River and Into the
    Trees

True at First Light

Khaled Hosseini—3 Novels

The Kite Runner

A Thousand Splendid Suns

And the Mountains Echoed

E L James—3 Fifty Shades Novels

Fifty Shades of Grey

Fifty Shades Darker

Fifty Shades Freed

Henry James—20 Novels

Watch and Ward

Roderick Hudson

The American

The Europeans

Confidence

Washington Square

The Portrait of a Lady

The Bostonians

The Princess Casamassima

The Reverberator

The Tragic Muse

The Other House

The Spoils of Poynton

What Maisie Knew

The Awkward Age

The Sacred Fount

The Wings of the Dove

The Ambassadors

The Golden Bowl

The Outcry

Edward P. Jones—3 Books

Lost in the City

The Known World

All Aunt Hagar's Children

James Joyce—3 Novels

Ulysses

Finnegans Wake

A Portrait of the Artist as a
    Young Man

Stephen King—51 Novels

Carrie
'Salem's Lot
The Shining
Rage
The Stand
The Long Walk
The Dead Zone
Firestarter
Roadwork
Cujo
The Running Man
The Dark Tower I: The Gunslinger
Christine
Pet Sematary
Cycle of the Werewolf
The Talisman
Thinner
It
The Eyes of the Dragon
The Dark Tower II: The Drawing of the Three
Misery
The Tommyknockers
The Dark Half
The Dark Tower III: The Waste Lands
Needful Things
Gerald's Game
Dolores Claiborne
Insomnia
Rose Madder
The Green Mile
Desperation
The Regulators
The Dark Tower IV: Wizard and Glass
Bag of Bones
The Girl Who Loved Tom Gordon
Dreamcatcher
From a Buick 8
The Dark Tower V: Wolves of the Calla
The Dark Tower VI: Song of Susannah
The Dark Tower VII: The Dark Tower
Colorado Kid
Cell
Lisey's Story
Blaze
Duma Key
Under the Dome
11/22/63
The Dark Tower: The Wind Through the Keyhole
Joyland
Doctor Sleep
Mr. Mercedes

Rudyard Kipling—3 Novels

The Light That Failed
Captains Courageous
Kim

D. H. Lawrence—12 Novels

The Plumed Serpent

Sons and Lovers

The Boy in the Bush

The Rainbow

Kangaroo

The Lost Girl

The White Peacock

The Escaped Cock

Aaron's Rod

The Trespasser

Lady Chatterley's Lover

Women in Love

Elmore Leonard—45 Novels

The Bounty Hunters

The Law at Randado

Escape from Five Shadows

Last Stand at Saber River

Hombre

The Big Bounce

The Moonshine War

Valdez Is Coming

Forty Lashes Less One

Mr. Majestyk

52 Pickup

Swag

Unknown Man #89

The Hunted

The Switch

Gunsights

City Primeval

Gold Coast

Split Images

Cat Chaser

La Brava

Stick

Glitz

Bandits

Touch

Freaky Deaky

Killshot

Get Shorty

Maximum Bob

Rum Punch

Pronto

Riding the Rap

Out of Sight

Cuba Libre

Be Cool

Pagan Babies

Tishomingo Blues

A Coyote's in the House

Mr. Paradise

The Hot Kid

Comfort to the Enemy

Up in Honey's Room

Road Dogs

Djibouti

Raylan

Ira Levin—7 Novels

A Kiss Before Dying

Rosemary's Baby

This Perfect Day

The Stepford Wives

The Boys from Brazil

Silver

Son of Rosemary

C. S. Lewis—7 Narnia Novels

The Magician's Nephew

*The Lion, the Witch and the
    Wardrobe*

The Horse and His Boy

Prince Caspian

The Voyage of the Dawn Treader

The Silver Chair

The Last Battle

Sinclair Lewis—19 Novels*

Main Street

Babbitt

Free Air

Gideon Planish

Arrowsmith

Bethel Merriday

Elmer Gantry

The Job

Kingsblood Royal

The Prodigal Parents

Work of Art

Cass Timberlane

The Trail of the Hawk

The Innocents

Our Mr. Wrenn

It Can't Happen Here

World So Wide

Dodsworth

The God-Seeker

Jack London—20 Novels

The Cruise of the Dazzler

A Daughter of the Snows

The Call of the Wild

The Sea-Wolf

The Game

White Fang

Before Adam

The Iron Heel

Martin Eden

Burning Daylight

Adventure

The Scarlet Plague

A Son of the Sun

The Abysmal Brute

The Valley of the Moon

The Mutiny of the Elsinore

The Star Rover

The Little Lady of the Big House

Jerry of the Islands

Michael, Brother of Jerry

Lois Lowry—4 Novels

The Giver

Gathering Blue

Messenger

Son

---

* Not all of Sinclair Lewis's novels exist in digital form. The 19 listed here were
used. Four Lewis novels are missing: *Mantrap*, *The Man Who Knew Coolidge*,
Ann Vickers, and *Harri*.

George R. R. Martin—8 Novels

Dying of the Light	*A Clash of Kings*
Fevre Dream	*A Storm of Swords*
The Armageddon Rag	*A Feast for Crows*
A Game of Thrones	*A Dance with Dragons*

Cormac McCarthy—10 Novels

The Orchard Keeper	*All the Pretty Horses*
Outer Dark	*The Crossing*
Child of God	*Cities of the Plain*
Suttree	*No Country for Old Men*
Blood Meridian	*The Road*

Ian McEwan—13 Novels

The Cement Garden	*Atonement*
The Comfort of Strangers	*Saturday*
The Child in Time	*On Chesil Beach*
The Innocent	*Solar*
Black Dogs	*Sweet Tooth*
Enduring Love	*The Children Act*
Amsterdam	

Richelle Mead—23 Novels

Succubus Blues	*Vampire Academy*
Succubus on Top	*Frostbite*
Succubus Dreams	*Shadow Kiss*
Succubus Heat	*Blood Promise*
Succubus Shadows	*Spirit Bound*
Succubus Revealed	*Last Sacrifice*
Storm Born	*Bloodlines*
Thorn Queen	*The Golden Lily*
Iron Crowned	*The Indigo Spell*
Shadow Heir	*The Fiery Heart*
Gameboard of the Gods	*Silver Shadows*
The Immortal Crown	

Herman Melville—9 Novels

Typee

Omoo

Mardi

Redburn

White-Jacket

Moby Dick

Pierre

Israel Potter

The Confidence-Man

Stephenie Meyer—4 Twilight Novels

Twilight

New Moon

Eclipse

Breaking Dawn

David Mitchell—6 Novels

Ghostwritten

number9dream

Cloud Atlas

Black Swan Green

The Thousand Autumns of Jacob
  de Zoet

The Bone Clocks

Toni Morrison—10 Novels

Beloved

Jazz

Love

Song of Solomon

Tar Baby

Paradise

Home

A Mercy

Sula

The Bluest Eye

Vladimir Nabokov—8 Novels

Ada, or Ardor

Bend Sinister

Lolita

Look at the Harlequins!

Pale Fire

Pnin

The Real Life of Sebastian Knight

Transparent Things

George Orwell—6 Novels

Coming Up for Air

Animal Farm

Burmese Days

Keep the Aspidistra Flying

1984

A Clergyman's Daughter

Chuck Palahniuk—14 Novels

Fight Club

Survivor

Invisible Monsters

Choke

Lullaby

Diary

Haunted

Rant

Snuff

Pygmy

Tell-All

Damned

Doomed

Beautiful You

James Patterson—22 Alex Cross Novels

Along Came a Spider

Kiss the Girls

Jack & Jill

Cat & Mouse

Pop Goes the Weasel

Roses Are Red

Violets Are Blue

Four Blind Mice

The Big Bad Wolf

London Bridges

Mary, Mary

Cross

Double Cross

Cross Country

Alex Cross's Trial

I, Alex Cross

Cross Fire

Kill Alex Cross

Merry Christmas, Alex Cross

Alex Cross, Run

Cross My Heart

Hope to Die

Louise Penny—10 Novels

Still Life

A Fatal Grace

The Cruelest Month

The Murder Stone

Brutal Telling

Bury Your Dead

A Trick of the Light

The Beautiful Mystery

How the Light Gets In

The Long Way Home

Jodi Picoult—21 Novels

Songs of the Humpback Whale
Harvesting the Heart
Picture Perfect
Mercy
The Pact
Keeping Faith
Plain Truth
Salem Falls
Perfect Match
Second Glance
My Sister's Keeper

Vanishing Acts
The Tenth Circle
Nineteen Minutes
Change of Heart
Handle with Care
House Rules
Sing You Home
Lone Wolf
The Storyteller
Leaving Time

Thomas Pynchon—8 Novels

V
The Crying of Lot 49
Gravity's Rainbow
Vineland

Mason & Dixon
Against the Day
Inherent Vice
Bleeding Edge

Ayn Rand—3 Novels

We the Living
Atlas Shrugged

The Fountainhead

Rick Riordan—5 Percy Jackson Novels

The Lightning Thief
The Sea of Monsters
The Titan's Curse

The Battle of Labyrinth
The Last Olympian

Marilynne Robinson—4 Novels

Housekeeping
Gilead

Home
Lila

Veronica Roth—3 Divergent Novels

Divergent
Allegiant

Insurgent

J. K. Rowling—7 Harry Potter Novels

Harry Potter and the Sorcerer's Stone

Harry Potter and the Chamber of Secrets

Harry Potter and the Prisoner of Azkaban

Harry Potter and the Goblet of Fire

Harry Potter and the Order of the Phoenix

Harry Potter and the Half-Blood Prince

Harry Potter and the Deathly Hallows

Salman Rushdie—9 Novels

Grimus

Midnight's Children

Shame

The Satanic Verses

The Moor's Last Sigh

The Ground Beneath Her Feet

Fury

Shalimar the Clown

The Enchantress of Florence

Alice Sebold—3 Novels

Lucky

The Lovely Bones

The Almost Moon

Zadie Smith—4 Novels

White Teeth

The Autograph Man

On Beauty

NW

Lemony Snicket—13 Unfortunate Events Novels

The Bad Beginning

The Reptile Room

The Wide Window

The Miserable Mill

The Austere Academy

The Ersatz Elevator

The Vile Village

The Hostile Hospital

The Carnivorous Carnival

The Slippery Slope

The Grim Grotto

The Penultimate Peril

The End

Nicholas Sparks—18 Novels

The Notebook

Message in a Bottle

A Walk to Remember

The Rescue

A Bend in the Road

Nights in Rodanthe

The Guardian

The Wedding

Three Weeks with My Brother

True Believer

At First Sight

Dear John

The Choice

The Lucky One

The Last Song

Safe Haven

The Best of Me

The Longest Ride

Danielle Steel—92 Novels

Going Home

Passion's Promise

Now and Forever

The Promise

Season of Passion

Summer's End

The Ring

Palomino

To Love Again

Remembrance

Loving

Once in a Lifetime

Crossings

A Perfect Stranger

Thurston House

Changes

Full Circle

Family Album

Secrets

Wanderlust

Fine Things

Kaleidoscope

Zoya

Star

Daddy

Message from Nam

Heartbeat

No Greater Love

Jewels

Mixed Blessings

Vanished

Accident

The Gift

Wings

Lightning

Five Days in Paris

Malice

Silent Honor

The Ranch

Special Delivery

The Ghost

The Long Road Home

The Klone and I

Mirror Image

Bittersweet

Granny Dan

Irresistible Forces

The Wedding

The House on Hope Street
Journey
Lone Eagle
Leap of Faith
The Kiss
The Cottage
Sunset in St. Tropez
Answered Prayers
Dating Game
Johnny Angel
Safe Harbour
Ransom
Second Chance
Echoes
Impossible
Miracle
Toxic Bachelors
The House
Coming Out
H.R.H.
Sisters
Bungalow 2

Amazing Grace
Honor Thyself
Rogue
A Good Woman
One Day at a Time
Matters of the Heart
Southern Lights
Big Girl
Family Ties
Legacy
44 Charles Street
Happy Birthday
Hotel Vendôme
Betrayal
Friends Forever
The Sins of the Mother
Until the End of Time
First Sight
Winners
Power Play
A Perfect Life
Pegasus

John Steinbeck—19 Novels

The Winter of Our Discontent
In Dubious Battle
The Grapes of Wrath
Sweet Thursday
Of Mice and Men
East of Eden
The Pearl
The Wayward Bus
Travels with Charley
The Acts of King Arthur and His
   Noble Knights

Burning Bright
Cannery Row
Tortilla Flat
The Short Reign of Pippin IV
Cup of Gold
The Red Pony
Bombs Away
To a God Unknown
The Moon Is Down

R.L. Stine—62 Goosebumps Novels

Welcome to Dead House
Stay Out of the Basement
Monster Blood
Say Cheese and Die!
The Curse of the Mummy's Tomb
Let's Get Invisible
Night of the Living Dummy
The Girl Who Cried Monster
Welcome to Camp Nightmare
The Ghost Next Door
The Haunted Mask
Be Careful What You Wish For . . .
Piano Lessons Can Be Murder
The Werewolf of Fever Swamp
You Can't Scare Me!
One Day at Horrorland
Why I'm Afraid of Bees
Monster Blood II
Deep Trouble
The Scarecrow Walks at Midnight
Go Eat Worms!
Ghost Beach
Return of the Mummy
Phantom of the Auditorium
Attack of the Mutant
My Hairiest Adventure
A Night in Terror Tower
The Cuckoo Clock of Doom
Monster Blood III
It Came from Beneath the Sink!
Night of the Living Dummy II
The Barking Ghost

The Horror at Camp Jellyjam
Revenge of the Lawn Gnomes
A Shocker on Shock Street
The Haunted Mask II
The Headless Ghost
The Abominable Snowman of
    Pasadena
How I Got My Shrunken Head
Night of the Living Dummy III
Bad Hare Day
Egg Monsters from Mars
The Beast from the East
Say Cheese and Die—Again!
Ghost Camp
How to Kill a Monster
Legend of the Lost Legend
Attack of the Jack-O'-Lanterns
Vampire Breath
Calling All Creeps!
Beware, the Snowman
How I Learned to Fly
Chicken, Chicken
Don't Go to Sleep!
The Blob That Ate Everyone
The Curse of Camp Cold Lake
My Best Friend Is Invisible
Deep Trouble II
The Haunted School
Werewolf Skin
I Live in Your Basement
Monster Blood IV

Amy Tan—6 Novels

The Joy Luck Club
The Kitchen God's Wife
The Hundred Secret Senses

The Bonesetter's Daughter
Saving Fish from Drowning
The Valley of Amazement

Donna Tartt—3 Novels

The Secret History
The Little Friend

The Goldfinch

J. R. R. Tolkien—4 Novels

The Fellowship of the Ring
The Two Towers

The Return of the King
The Hobbit

Mark Twain—13 Books

The Gilded Age
The Adventures of Tom Sawyer
The Prince and the Pauper
*The Adventures of Huckleberry
    Finn*
*A Connecticut Yankee in King
    Arthur's Court*
The American Claimant
Tom Sawyer Abroad

Pudd'nhead Wilson
Tom Sawyer, Detective
*Personal Recollections of Joan
    of Arc*
A Double Barrelled Detective Story
    (novella)
A Horse's Tale (novella)
The Mysterious Stranger

John Updike—26 Novels

Rabbit, Run
Rabbit Redux
Rabbit Is Rich
Rabbit at Rest
Bech: A Book
Bech Is Back
Bech at Bay
*Memories of the Ford
    Administration*
The Witches of Eastwick
The Widows of Eastwick
A Month of Sundays
Roger's Version
S

The Poorhouse Fair
The Centaur
Of the Farm
Couples
Marry Me
The Coup
Brazil
In the Beauty of the Lilies
Toward the End of Time
Gertrude and Claudius
Seek My Face
Village
Terrorist

Kurt Vonnegut—14 Novels

Timequake

Slaughterhouse-Five

Slapstick

The Sirens of Titan

Player Piano

Mother Night

Jailbird

Hocus Pocus

God Bless You, Mr. Rosewater

Galápagos

Deadeye Dick

Cat's Cradle

Breakfast of Champions

Bluebeard

Alice Walker—8 Novels

The Third Life of Grange Copeland

Meridian

The Color Purple

The Temple of My Familiar

Possessing the Secret of Joy

By the Light of My Father's Smile

The Way Forward Is with a Broken
   Heart

Now Is the Time to Open Your
   Heart

Edith Wharton—22 Novels

Ethan Frome

The Valley of Decision

Bunner Sisters

Summer

Sanctuary

The Gods Arrive

The Touchstone

Hudson River Bracketed

The Custom of the Country

The Marne: A Tale of War

A Son at the Front

The Reef

The Fruit of the Tree

The Age of Innocence

The Buccaneers

The Glimpses of the Moon

Twilight Sleep

The Mother's Recompense

The Children

Fast and Loose

Madame de Treymes

The House of Mirth

E. B. White—3 Novels

Charlotte's Web

Stuart Little

The Trumpet of the Swan

Tom Wolfe—4 Novels

A Man in Full

Back to Blood

I Am Charlotte Simmons

The Bonfire of the Vanities

Virginia Woolf—9 Novels

The Voyage Out	*Orlando*
Night and Day	*The Waves*
Jacob's Room	*The Years*
Mrs. Dalloway	*Between the Acts*
To the Lighthouse	

Markus Zusak—5 Novels

The Underdog	*The Messenger*
Fighting Ruben Wolfe	*The Book Thief*
Getting the Girl	

Additional Bibliography Lists Used
(in order of appearance)

Adverb Chart for 15 Authors—Introduced Chapter 1

The authors chosen for this early graph were picked to represent a variety of time periods and genres before moving on to the "Great Books" list that dominates the rest of the chapter. I wanted to start with a smaller sample before moving to the full fifty-author list. All other lists in the chapter, including the "Great Books by Great Authors" list, are delineated within the text.

Goodreads Data: Average Rating vs. Number of Ratings— Introduced Chapter 1

The huge sample size afforded by Goodreads helps overcome the idiosyncrasies of any particular book critic, who might like one book over another for personal or peculiar reasons. But using the numerical rating itself (e.g., 3.5 stars out of 5.0) is still imperfect; and instead the *number* of ratings a book has been given ends up being a better approximation of the book's quality. The reason that the average rating falters as a metric is that the sample size can differ drastically from book to book, and this ends up skewing the rating in surprising ways.

To see the sampling issues that arise in a rating system, consider the numerical ratings for the Twilight series. The first book in the tetralogy has over 2.3 million ratings and averages a 3.56 out of 5. The fourth book has far fewer ratings (over 850,000), but averages a 3.72. The first book had 23% of readers give it just one or two stars. One could presume that the people who did not like the first book would not keep reading the series at the same rate as those who loved it. As a result, those who already liked the series kept reading and the average satisfaction derived from the books rose from their continual high ratings.

This selection bias is even more obvious when looking at the ratings for *Midnight Sun*. In 2008, an unfinished version of what was supposed to be a retelling of the first book in Meyer's series (but from the perspective of vampire Edward Cullen) was leaked online. This book was never finished, edited, or promoted, but on Goodreads.com it has a higher rating, 4.06, than any of the Twilight books. It does, however, have fewer ratings, around 125,000. These 125,000 are people who went far out of their way to read the text, meaning the large majority of them could be considered superfans. These are readers who already loved Stephenie Meyer, and already loved Edward Cullen, which helps explain why an unreleased book has such a high rating.

Many other series follow this same pattern. *Harry Potter and the Deathly Hallows*, the seventh and final book in the Harry Potter series, has half the ratings of the first Harry Potter book, but is rated the highest in the series while the first book has the second-lowest rating. Each of the three books in the *Lord of the Rings* trilogy increases in rating while the number of ratings they receive goes down. The three books in Asimov's Foundation trilogy follow this exact same pattern.

So what does any of this have to do with adverbs? It's an illustration that finding an objective measure of a book is not a straightforward task. *The Pearl* has more than 97,000 ratings on

Goodreads.com with a rating of 3.33. Meanwhile, Steinbeck's novel *To a God Unknown* has a rating of 3.90 with just 4,200 ratings. Many more people have been in an English class where they had to read *The Pearl* whether they wanted to or not, while select fans waded through several other Steinbeck books first to get to the latter. *To a God Unknown* even has a higher rating than *The Grapes of Wrath* and *Of Mice and Men*, even though these books have hundreds of thousands more reviews and are widely considered Steinbeck's greatest works. The same pattern can be seen on Amazon as well. Much like a sequel whittling down to an author's true fan base, the smaller sample size skews the rating when we compare an author's more obscure books to their classics. And for this reason, the total number of ratings received by a book becomes a better approximation of its quality and status.

Classic Literature—Introduced Chapter 2

This list consists of the top fifty books written by men and the top fifty books written by women on the composite list "The Best English-Language Fiction of the Twentieth Century" assembled by Brian Kunde.

A Thousand Acres—Jane Smiley
A Tree Grows in Brooklyn—Betty Smith
The Age of Innocence—Edith Wharton
Bastard Out of Carolina—Dorothy Allison
Beloved—Toni Morrison
Burger's Daughter—Nadine Gordimer
Cold Sassy Tree—Olive Ann Burns
Death Comes for the Archbishop—Willa Cather
Ellen Foster—Kaye Gibbons
Ethan Frome—Edith Wharton
Gone with the Wind—Margaret Mitchell
A Good Man Is Hard to Find—Flannery O'Connor
Jazz—Toni Morrison
Little House in the Big Woods—Laura Ingalls Wilder
Rebecca—Daphne du Maurier

Mrs. Dalloway—Virginia Woolf
My Ántonia—Willa Cather
O Pioneers!—Willa Cather
Ordinary People— Judith Guest
Orlando—Virginia Woolf
Pale Horse, Pale Rider—Katherine Anne Porter
Possession—A. S. Byatt
Atlas Shrugged—Ayn Rand
Rubyfruit Jungle—Rita Mae Brown
Song of Solomon—Toni Morrison
Sula—Toni Morrison
Talk Before Sleep— Elizabeth Berg
The Bean Trees—Barbara Kingsolver
The Bell Jar—Sylvia Plath
The Bluest Eye—Toni Morrison
The Color Purple—Alice Walker
The Death of the Heart—Elizabeth Bowen
The Fountainhead—Ayn Rand
The Golden Notebook—Doris Lessing
The Good Earth—Pearl Buck
The Handmaid's Tale—Margaret Atwood
The Heart Is a Lonely Hunter—Carson McCullers
The House of Mirth—Edith Wharton
The Joy Luck Club—Amy Tan
The Mists of Avalon—Marion Bradley
The Prime of Miss Jean Brodie—Muriel Spark
The Shell Seekers—Rosamunde Pilcher
The Shipping News—Annie Proulx
The Stone Diaries—Carol Shields
The Thorn Birds—Colleen McCullough
Their Eyes Were Watching God—Zora Neale Hurston
To Kill a Mockingbird—Harper Lee
To the Lighthouse—Virginia Woolf
Under the Net—Iris Murdoch
Wide Sargasso Sea—Jean Rhys

A Clockwork Orange—Anthony Burgess
A Farewell to Arms—Ernest Hemingway

A Passage to India—E. M. Forster
A Portrait of the Artist as a Young Man—James Joyce
A Room with a View—E. M. Forster
All the King's Men—Robert Penn Warren
An American Tragedy—Theodore Dreiser
Animal Farm—George Orwell
As I Lay Dying—William Faulkner
Brave New World—Aldous Huxley
Catch-22—Joseph Heller
Charlotte's Web—E. B. White
Fahrenheit 451—Ray Bradbury
For Whom the Bell Tolls—Ernest Hemingway
Go Tell It on the Mountain—James Baldwin
Heart of Darkness—Joseph Conrad
Howards End—E. M. Forster
I, Claudius—Robert Graves
Invisible Man—Ralph Ellison
Lady Chatterley's Lover—D. H. Lawrence
Lolita—Vladimir Nabokov
Lord of the Flies—William Golding
Lord of the Rings (all)—J. R. R. Tolkien
Native Son—Richard Wright
1984—George Orwell
Of Mice and Men—John Steinbeck
On the Road—Jack Kerouac
One Flew Over the Cuckoo's Nest—Ken Kesey
Slaughterhouse-Five—Kurt Vonnegut
Sons and Lovers—D. H. Lawrence
Sophie's Choice—William Styron
Stranger in a Strange Land—Robert A. Heinlein
Tender Is the Night—F. Scott Fitzgerald
The Call of the Wild—Jack London
The Catcher in the Rye—J. D. Salinger
The Good Soldier—Ford Madox Ford
The Grapes of Wrath—John Steinbeck
The Great Gatsby—F. Scott Fitzgerald
The Hobbit—J. R. R. Tolkien
The Jungle—Upton Sinclair

The Old Man and the Sea—Ernest Hemingway
The Sound and the Fury—William Faulkner
The Sun Also Rises—Ernest Hemingway
The Way of All Flesh—Samuel Butler
The Wings of the Dove—Henry James
The Wonderful Wizard of Oz—L. Frank Baum
The World According to Garp—John Irving
Ulysses—James Joyce
Winesburg, Ohio—Sherwood Anderson
Winnie the Pooh—A. A. Milne

Modern Popular Fiction—Introduced Chapter 2

Starting with the end of 2014 and going backward, this list consists of the last fifty number one *New York Times* fiction bestsellers written by women and the last fifty by men. No books with co-authors were included.

A Week in Winter—Maeve Binchy
Believing the Lie—Elizabeth George
Big Little Lies—Liane Moriarty
Calculated in Death—J. D. Robb
Celebrity in Death—J. D. Robb
Chasing Fire—Nora Roberts
Daddy's Gone a Hunting—Mary Higgins Clark
Dead Ever After—Charlaine Harris
Dead Reckoning—Charlaine Harris
Deadlocked—Charlaine Harris
Dreams of Joy—Lisa See
Concealed in Death—J. D. Robb
Explosive Eighteen—Janet Evanovich
Flash and Bones—Kathy Reichs
Frost Burned—Patricia Briggs
Gone Girl—Gillian Flynn
Hit List—Laurell K. Hamilton
Home Front—Kristin Hannah
How the Light Gets In—Louise Penny
I've Got You Under My Skin—Mary Higgins Clark

Kiss the Dead—Laurell K. Hamilton
Leaving Time—Jodi Picoult
Lone Wolf—Jodi Picoult
Lover at Last—J. R. Ward
Lover Reborn—J. R. Ward
New York to Dallas—J. D. Robb
Notorious Nineteen—Janet Evanovich
Power Play—Danielle Steel
Shadow of Night—Deborah Harkness
Smokin' Seventeen—Janet Evanovich
Starting Now—Debbie Macomber
Takedown Twenty—Janet Evanovich
The Book of Life—Deborah Harkness
The Casual Vacancy—J. K. Rowling
The Collector—Nora Roberts
The Cuckoo's Calling—Robert Galbraith
The Goldfinch—Donna Tartt
The Invention of Wings—Sue Monk Kidd
The King—J. R. Ward
The Long Way Home—Louise Penny
The Lost Years—Mary Higgins Clark
The One & Only—Emily Giffin
The Storyteller—Jodi Picoult
The Undead Pool—Kim Harrison
Top Secret Twenty-One—Janet Evanovich
Until the End of Time—Danielle Steel
W Is for Wasted—Sue Grafton
Whiskey Beach—Nora Roberts
Wicked Business—Janet Evanovich
Written in My Own Heart's Blood—Diana Gabaldon

11-22-63—Stephen King
77 Shadow Street—Dean Koontz
A Wanted Man—Lee Child
Act of War—Brad Thor
Alex Cross, Run—James Patterson
Calico Joe—John Grisham

Cold Days—Jim Butcher
Colorless Tsukuru Tazaki and His Years of Pilgrimage—
  Haruki Murakami
Cross My Heart—James Patterson
Deadline—John Sandford
Doctor Sleep—Stephen King
Edge of Eternity—Ken Follett
Gray Mountain—John Grisham
Hope to Die—James Patterson
Inferno—Dan Brown
Kill Alex Cross—James Patterson
Kill Shot—Vince Flynn
Missing You—Harlan Coben
Mr. Mercedes—Stephen King
Never Go Back—Lee Child
Personal—Lee Child
Revival—Stephen King
Six Years—Harlan Coben
Skin Game—Jim Butcher
Stay Close—Harlan Coben
Stolen Prey—John Sandford
Storm Front—John Sandford
Sycamore Row—John Grisham
Taken—Robert Crais
The Drop—Michael Connelly
The English Girl—Daniel Silva
The Escape—David Baldacci
The Fallen Angel—Daniel Silva
The First Phone Call from Heaven—Mitch Albom
The Heist—Daniel Silva
The Hit—David Baldacci
The Innocent—David Baldacci
The Last Man—Vince Flynn
The Litigators—John Grisham
The Longest Ride—Nicholas Sparks
The Magicians—Lev Grossman
The Ocean at the End of the Lane—Neil Gaiman
The Panther—Nelson DeMille

The Racketeer—John Grisham
The Target—David Baldacci
The Time Keeper—Mitch Albom
The Wind Through the Keyhole—Stephen King
Winter of the World—Ken Follett
Words of Radiance—Brandon Sanderson
Zero Day—David Baldacci

Modern Literary Fiction—Introduced Chapter 2

Starting with awards given at the end of 2014 and looking backward, this list consists of the last fifty novels written by women and the last fifty novels written by men that were on any of the following lists: *New York Times* Top Ten Books of the Year, Pulitzer Prize finalists, Man Booker Prize short list, National Book Award finalists, National Book Critics Circle finalists, and *Time* magazine's best books of the year. I did not exclude a book from this list even if it had been a bestseller (for example Stephen King's *11-22-63*).

A Gate at the Stairs—Lorrie Moore
A Tale for the Time Being—Ruth Ozeki
A Visit from the Goon Squad—Jennifer Egan
Americanah—Chimamanda Ngozi Adichie
Bring Up the Bodies—Hilary Mantel
Dept. of Speculation—Jenny Offill
Euphoria—Lily King
Faithful Place—Tana French
Great House—Nicole Krauss
Half Broke Horses—Jeannette Walls
Half-Blood Blues—Esi Edugyan
How to Be Both—Ali Smith
Jamrach's Menagerie—Carol Birch
Lark and Termite—Jayne Anne Phillips
Life After Life—Kate Atkinson
Lila—Marilynne Robinson
Lord of Misrule—Jaimy Gordon
Love in Infant Monkeys—Lydia Millet
Magnificence—Lydia Millet

At Last—Edward St. Aubyn
Billy Lynn's Long Halftime Walk—Ben Fountain
Bleeding Edge—Thomas Pynchon
C—Tom McCarthy
Family Life—Akhil Sharma
Freedom—Jonathan Franzen
Harvest—Jim Crace
In a Strange Room—Damon Galgut
J—Howard Jacobson
Narcopolis—Jeet Thayil
NOS4A2—Joe Hill
On Such a Full Sea—Chang-rae Lee
Open City—Teju Cole
Parrot and Olivier in America—Peter Carey
Pigeon English—Stephen Kelman
Snowdrops—A. D. Miller
The Art of Fielding—Chad Harbach
The Bone Clocks—David Mitchell
The Family Fang—Kevin Wilson
The Fault in Our Stars—John Green
The Garden of Evening Mists—Tan Twan Eng
The Good Lord Bird—James McBride
The Infatuations—Javier Marías
The Laughing Monsters—Denis Johnson
The Lives of Others—Neel Mukherjee
The Marriage Plot—Jeffrey Eugenides
The Narrow Road to the Deep North—Richard Flanagan
The Ocean at the End of the Lane—Neil Gaiman
The Orphan Master's Son—Adam Johnson
The Pale King—David Foster Wallace
The Privileges—Jonathan Dee
The Sense of an Ending—Julian Barnes
The Sisters Brothers—Patrick Dewitt
The Sojourn—Andrew Krivak
The Son—Philipp Meyer
The Stranger's Child—Alan Hollinghurst
The Surrendered—Chang-rae Lee
The Testament of Mary—Colm Tóibín

The Woman Who Lost Her Soul—Bob Shacochis

The Zone of Interest—Martin Amis

To Rise Again at a Decent Hour—Joshua Ferris

Train Dreams—Denis Johnson

Umbrella—Will Self

Yellow Birds—Kevin Powers

Male/Female Indicative Words—Introduced Chapter 2

The sources for the chart on page 33 are as follows:

1. *Facebook Status data*: H. A. Schwartz, J. C. Eichstaedt, M. L. Kern, L. Dziurzynski, S. M. Ramones, M. Agrawal, et al. "Personality, Gender, and Age in the Language of Social Media: The Open-Vocabulary Approach," Public Library of Science One 8(9): e73791. 2, 2013.

2. *Chatroom Emoticons data*: S. Kapidzic, S. C. Herring. "Gender, Communication, and Self-presentation in Teen Chatrooms Revisited: Have Patterns Changed?" *Journal of Computer-Mediated Communication*, 2011: 17, 39–59.

3. *Twitter Assent or Negation Terms data*: D. Bamman, J. Eisenstein, and T. Schnoebelen. "Gender Identity and Lexical Variation in Social Media," *Journal of Sociolinguistics*, 18: 135–160, doi: 10.1111/josl.12080.

4. *Blogs data*: J. Schler, M. Koppel, S. Argamon, and J. Pennebaker. "Effects of Age and Gender on Blogging," *Computational Approaches to Analyzing Weblogs—Papers from the AAAI Spring Symposium, Technical Report*, 2006, vol. SS-06-03, pp. 191–197.

Starting on page 37, I also discuss a method created by Neal Krawetz for guessing the gender of an author. It is a more basic version of the academic paper, and I chose to include it for its simplicity. Each word is weighted based on how common it is in the writing of one gender compared to the other, and I combined the scores of Krawetz's "formal" and "informal" method. Also, as mentioned in the text, gendered pronouns were removed from the scoring system so that pronouns like *she* and *he* alone were not giving away results.

The point values in the original method developed by Krawetz were scaled so the average sample would have a 1:1 ratio of male-weighted words to female-weighted words. Because I removed gendered pronouns, which are considered indicative of female writing, the ratio of male word points to female word points was closer to 6:5 in the three samples. To account for this, I rescaled the scores so that the point ratio was 1:1 over the three samples.

If I kept Krawetz's original method intact (leaving in all gendered pronouns and not changing the scaling) the results would have correctly predicted 63, 71, and 59 of all books in the classic, modern bestseller, and modern literary samples respectively. This is better than the 58, 66, and 58 discussed in the text. However, because I chose to examine if nongendered words have correlation to the gender of the author in popular literature, the scaling was necessary. Krawetz's method is available at www.hackerfactor.com/GenderGuesser under the "Genre: Formal" section.

Below are the 47 words and their point value according to the Krawetz method. The points shown for male words are the points given by the original method. The points shown for female words have been scaled by roughly 1.19 for the purposes of Chapter 2.

Male Indicative Words (according to Krawetz's method)

a +6	*in +10*
above +4	*is +18*
are +28	*it +6*
around +42	*many + 6*
as +60	*now +33*
at +6	*said +5*
below +8	*some +58*
ever +21	*something +26*
good +31	*the +24*

these + 8	well +15
this +44	what +35
to +2	who +19

Female Indicative Words (according to Krawetz's method)

actually +49	out +39
am +42	should +7
and +4	since +25
be +17	so +54
because +55	too +38
but +43	was +1
everything +44	we +8
has +33	when +17
if +22	where +18
like +43	with +52
more +7	your +19
not +27	

Fifty Author List—Introduced Chapter 3

The fifty authors included in this list were selected to represent a mix of literary fiction and bestsellers, both modern hits and classics. I chose from among this book's earlier lists while also making several additions (like Elmore Leonard) to cover a broader range of genres and time periods. This list is also used in later chapters of the book, but in Chapter 4 and beyond Harper Lee is included in place of Thomas Pynchon. Mosteller and Wallace's author identification method requires at least two works, a known sample and an unknown sample. However, when this list was assembled Harper Lee had published only one book. Thomas Pynchon was chosen as a replacement so as to investigate his onetime authorship controversy.

Jane Austen—6 novels	Agatha Christie—66 novels
Dan Brown—4 Robert Langdon books	Suzanne Collins—3 Hunger Games books
Willa Cather—14 novels	Joseph Conrad—14 novels
Michael Chabon—7 novels	Charles Dickens—20 novels

Theodore Dreiser—8 novels

Jennifer Egan—4 novels

Dave Eggers—6 novels

William Faulkner—19 novels

F. Scott Fitzgerald—4 novels

Gillian Flynn—3 novels

E. M. Forster—6 novels

Jonathan Franzen—4 novels

William Gaddis—5 novels

Neil Gaiman—7 novels

John Green—4 novels

Ernest Hemingway—10 novels

Khaled Hosseini—3 novels

E L James—3 Fifty Shades books

James Joyce—3 novels

Stephen King—51 novels

D. H. Lawrence—12 novels

Elmore Leonard—45 novels

Sinclair Lewis—19 novels

Jack London—20 novels

Stephenie Meyer—4 Twilight books

Toni Morrison—10 novels

Vladimir Nabokov—8 novels

George Orwell—6 novels

Chuck Palahniuk—14 novels

James Patterson—22 Alex Cross books

Thomas Pynchon—8 novels

Ayn Rand—3 novels

Veronica Roth—3 Divergent books

J. K. Rowling—7 Harry Potter books

Salman Rushdie—9 novels

Zadie Smith—4 novels

John Steinbeck—19 novels

J. R. R. Tolkien—*LOTR* and *The Hobbit*

Mark Twain—13 novels

John Updike—26 novels

Kurt Vonnegut—14 novels

Alice Walker—8 novels

Edith Wharton—22 novels

E. B. White—3 novels

Tom Wolfe—4 novels

Virginia Woolf—9 novels

Pulitzer Prize Winners—Introduced Chapter 5

Throughout this book I referred to a collection of Pulitzer Prize–winning novels. A listing of these can be found at www.pulitzer.org. Unless noted for a particular study, only the winner for each year's "Pulitzer Prize for Fiction" was examined (and not finalists). The span of years considered often varied depending on the particular study, but is specified within the main text. Some years do not have winners.

***New York Times* Number One Bestsellers—Introduced Chapter 5**

The section on the decline in reading level relied heavily on *New York Times* Number One Bestsellers between 1960 and 2014. The listings were accessed from www.hawes.com. Over the years the criteria for which books are considered for the *New York Times* bestseller list has changed. For the statistics in this book the hardcover rankings were used.

In later sections the same listings from www.hawes.com were used. In some cases, such as author nationality in Chapter 6, all bestsellers were included instead of just the number one bestsellers. This is noted in the text.

New York Times Bestsellers—Introduced Chapter 6

The section on U.K. bestsellers and author nationality relied on the weekly top ten bestsellers in *The New York Sunday Times*. Books were taken from the fiction "Hardcover" list. As the text states, only the years 1974, 1984, 1994, 2004, and 2014 were included in the statistics. A full analysis of a forty-year time period was unfeasible due to intermittent gaps in the data, but the years in between showed a similar pattern to those discussed in the text (which is to say that the years discussed are not outliers).

Publishers Weekly Top Ten—Introduced Chapter 7

This sample, in which James Patterson came out as the most clichéd, was composed of *Publishers Weekly* top ten bestselling novels of the year. The years 2000 through 2013 were considered. While normally this would be 140 books, 13 books were repeats from previous years or were excluded by myself for being novels targeted at a much younger age range (such as *Diary of a Wimpy Kid*). The 127 books considered are following:

The Brethren—John Grisham
The Mark: The Beast Rules the World— Tim LaHaye
    and Jerry B. Jenkins
The Bear and the Dragon—Tom Clancy
The Indwelling: The Beast Takes Possession—Tim LaHaye
    and Jerry B. Jenkins

The Last Precinct—Patricia Cornwell
Journey—Danielle Steel
The Rescue—Nicholas Sparks
Roses Are Red—James Patterson
Cradle and All—James Patterson
The House on Hope Street—Danielle Steel
Desecration—Tim LaHaye and Jerry B. Jenkins
Skipping Christmas—John Grisham
A Painted House—John Grisham
Dreamcatcher—Stephen King
The Corrections—Jonathan Franzen
Black House—Stephen King and Peter Straub
Last Man Standing—David Baldacci
Valhalla Rising—Clive Cussler
A Day Late and a Dollar Short—Terry McMillan
Violets Are Blue—James Patterson
Blindsighted—Karin Slaughter
The Summons—John Grisham
Red Rabbit—Tom Clancy
The Remnant—Jerry B. Jenkins and Tim LaHaye
The Lovely Bones—Alice Sebold
Prey—Michael Crichton
The Shelters of Stone—Jean M. Auel
Four Blind Mice—James Patterson
Everything's Eventual: 14 Dark Tales—Stephen King
The Nanny Diaries—Emma McLaughlin and Nicola Kraus
Harry Potter and the Order of the Phoenix—J. K. Rowling
The Da Vinci Code—Dan Brown
The Five People You Meet in Heaven—Mitch Albom
The King of Torts—John Grisham
Bleachers—John Grisham
Armageddon—Tim LaHaye and Jerry B. Jenkins
The Teeth of the Tiger—Tom Clancy
The Big Bad Wolf—James Patterson
Blow Fly—Patricia Cornwell
The Last Juror—John Grisham
Glorious Appearing—Tim LaHaye and Jerry B. Jenkins
Angels & Demons—Dan Brown

State of Fear—Michael Crichton
London Bridges—James Patterson
Trace—Patricia Cornwell
The Rule of Four—Ian Caldwell and Dustin Thomason
The Broker—John Grisham
Mary, Mary—James Patterson
At First Sight—Nicholas Sparks
Predator—Patricia Cornwell
True Believer—Nicholas Sparks
Light from Heaven—Jan Karon
The Historian—Elizabeth Kostova
The Mermaid Chair—Sue Monk Kidd
Eleven on Top—Janet Evanovich
For One More Day—Mitch Albom
Cross—James Patterson
Dear John—Nicholas Sparks
Next—Michael Crichton
Hannibal Rising—Thomas Harris
Lisey's Story—Stephen King
Twelve Sharp—Janet Evanovich
Cell—Stephen King
Beach Road—James Patterson and Peter de Jonge
The 5th Horseman—James Patterson and Maxine Paetro
Harry Potter and the Deathly Hallows—J. K. Rowling
A Thousand Splendid Suns—Khaled Hosseini
Playing for Pizza—John Grisham
The Choice—Nicholas Sparks
Lean Mean Thirteen—Janet Evanovich
Plum Lovin'—Janet Evanovich
Book of the Dead—Patricia Cornwell
The Quickie—James Patterson and Michael Ledwidge
The 6th Target—James Patterson and Maxine Paetro
The Darkest Evening of the Year—Dean Koontz
The Appeal—John Grisham
The Story of Edgar Sawtelle—David Wroblewski
The Host—Stephenie Meyer
Cross Country—James Patterson
The Lucky One—Nicholas Sparks

Fearless Fourteen—Janet Evanovich

The Christmas Sweater—Glenn Beck

Scarpetta—Patricia Cornwell

Your Heart Belongs to Me—Dean Koontz

The Lost Symbol—Dan Brown

The Associate—John Grisham

I, Alex Cross—James Patterson

The Last Song—Nicholas Sparks

Ford County—John Grisham

Finger Lickin' Fifteen—Janet Evanovich

Under the Dome—Stephen King

Pirate Latitudes—Michael Crichton

The Girl Who Kicked the Hornet's Nest—Stieg Larsson

The Confession—John Grisham

The Help—Kathryn Stockett

Safe Haven—Nicholas Sparks

Dead or Alive—Tom Clancy with Grant Blackwood

Sizzling Sixteen—Janet Evanovich

Cross Fire—James Patterson

Freedom—Jonathan Franzen

Port Mortuary—Patricia Cornwell

Full Dark, No Stars—Stephen King

The Litigators—John Grisham

11/22/63—Stephen King

The Best of Me—Nicholas Sparks

Smokin' Seventeen—Janet Evanovich

A Dance with Dragons—George R. R. Martin

Explosive Eighteen—Janet Evanovich

Kill Alex Cross—James Patterson

Micro—Michael Crichton

Dead Reckoning—Charlaine Harris

Locked On—Tom Clancy and Mark Greaney

Fifty Shades of Grey—E L James

The Hunger Games—Suzanne Collins

Fifty Shades Darker—E L James

Fifty Shades Freed—E L James

Catching Fire—Suzanne Collins

Mockingjay—Suzanne Collins

The Mark of Athena—Rick Riordan
Gone Girl—Gillian Flynn
Inferno—Dan Brown
The House of Hades—Rick Riordan
Divergent—Veronica Roth
Sycamore Row—John Grisham
Allegiant—Veronica Roth
Doctor Sleep—Stephen King
The Fault in Our Stars—John Green

Breakout Debut Novels—Introduced Chapter 8

A book was included in the sample if: (1) It was an author's first novel, (2) it was published between 1980 and 2014, and (3) it was a finalist for or winner of the Pulitzer Prize for Fiction, The National Book Award, or the National Book Critics Circle Award.

Birdy—William Wharton
Housekeeping—Marilynne Robinson
Leaving the Land—Douglas Unger
Jernigan—David Gates
Tinkers—Paul Harding
Vanished—Mary McGarry Morris
The Joy Luck Club—Amy Tan
Wartime Lies—Louis Begley
Dreaming in Cuban—Cristina Garcia
Cold Mountain—Charles Frazier
That Night—Alice McDermott
Three Junes—Julia Glass
Madeleine Is Sleeping—Sarah Shun-lien Bynum
Then We Came to the End—Joshua Ferris
Telex from Cuba—Rachel Kushner
During the Reign of the Queen of Persia—Joan Chase
Stones for Ibarra—Harriet Doerr
Love Medicine—Louise Erdrich
Three Farmers on Their Way to the Dance—Richard Powers
Typical American—Gish Jen
The Prince of West End Avenue—Alan Isler
White Teeth—Zadie Smith

Brick Lane—Monica Ali
In the Country of Men—Hisham Matar
The Ballad of Trenchmouth Taggart—M. Glenn Taylor

Hardy Boys—Introduced Chapter 9

The charts and statistics for the cliffhanger section in Chapter 9 came from the first seven books in the Hardy Boys series. In 1959 the series was revised, and all editions I used included these 1959 revisions.

The Tower Treasure
The House on the Cliff
The Secret of the Old Mill
The Missing Chums
Hunting for Hidden Gold
The Shore Road Mystery
The Secret of the Caves

Nancy Drew—Introduced Chapter 9

The charts and statistics for the cliffhanger section in Chapter 9 came from the first seven books in the Nancy Drew series. In 1959 the series was revised, and all editions I used included these 1959 revisions.

The Secret of the Old Clock
The Hidden Staircase
The Bungalow Mystery
The Mystery of Lilac Inn
The Secret of Shadow Ranch
The Secret of Red Gate Farm
The Clue in the Diary

About the Author

Ben Blatt is the co-author of *I Don't Care If We Never Get Back* and a former staff writer at *Slate* magazine. You can find him online at his website, BBlatt.com, or on Twitter @BenBlatt.